Aerodynamics of Low Reynolds Number Flyers

Low Reynolds number aerodynamics is important to a number of natural and man-made flyers. Birds, bats, and insects have been investigated by biologists for years, and active study in the aerospace engineering community, motivated by interest in micro air vehicles (MAVs), has been increasing rapidly. The primary focus of this book is the aerodynamics associated with fixed and flapping wings. The book considers both biological flyers and MAVs, including a summary of the scaling laws that relate the aerodynamics and flight characteristics to a flyer's sizing on the basis of simple geometric and dynamics analyses, structural flexibility, laminar–turbulent transition, airfoil shapes, and unsteady flapping-wing aerodynamics. The interplay between flapping kinematics and key dimensionless parameters such as the Reynolds number, Strouhal number, and reduced frequency is highlighted. The various unsteady lift-enhancement mechanisms are also addressed.

Wei Shyy is the Clarence L. "Kelly" Johnson Collegiate Professor and Chairman of the Department of Aerospace Engineering at the University of Michigan. He also taught at the University of Florida, as Distinguished Professor and Department Chair. He is the author and coauthor of books and articles dealing with computational and modeling techniques involving fluid flow, aerodynamics, propulsion, interfacial dynamics, and moving-boundary problems. He is the General Editor of the Cambridge Aerospace Series (Cambridge University Press), and is a Fellow of the American Institute of Aeronautics and Astronautics and the American Society of Mechanical Engineers.

Yongsheng Lian, Jian Tang, and Dragos Viieru are research scientists at the University of Michigan. They have done original research in flexible-wing and aerodynamics interactions: flapping-wing aerodynamics; laminar–turbulent transition; and unsteady, low Reynolds number fluid physics.

Hao Liu is a Professor of Biomechanical Engineering at Chiba University in Japan. He is well known for his contributions to biological, flapping-flight research, including original publications on insect aerodynamics simulations.

Cambridge Aerospace Series

Editors: Wei Shyy and Michael J. Rycroft

Aerodynamics of Low Reynolds Number Flyers

WEI SHYY

University of Michigan

YONGSHENG LIAN

University of Michigan

JIAN TANG

University of Michigan

DRAGOS VIIERU

University of Michigan

HAO LIU

Chiba University

CAMBRIDGE
UNIVERSITY PRESS

CAMBRIDGE UNIVERSITY PRESS
Cambridge, New York, Melbourne, Madrid, Cape Town,
Singapore, São Paulo, Delhi, Tokyo, Mexico City

Cambridge University Press
32 Avenue of the Americas, New York, NY 10013-2473, USA

www.cambridge.org
Information on this title: www.cambridge.org/9780521204019

First published 2008
Reprinted 2009
First paperback edition 2011

A catalog record for this publication is available from the British Library

Library of Congress Cataloging in Publication data

Shyy, Wei.
Aerodynamics of low reynolds number flyers : wei shyy . . . [et al.].
   p.   cm. – (Cambridge aerospace series)
Includes bibliographical references and index.
ISBN 978-0-521-88278-1 (hardback)
1. Aerodynamics.   1. Title.   11. Series.
TL570.S488   2007
629.132'3 – dc22   2007019227

ISBN 978-0-521-88278-1 Hardback
ISBN 978-0-521-20401-9 Paperback

Additional resources for this publication at www.cambridge.org/9780521204019

Contents

Nomenclature

k	reduced frequency	Eq. (1.1)
k	turbulent kinetic energy	Eq. (2.6)
l	characteristic length	Eq. (1.6)
L	lift	Eq. (1.3)
L_0	unstrained membrane length	Eq. (3.2)
L/D	lift-to-drag ratio, or glide ratio ($= C_L/C_D$)	Eq. (2.20)
m	body mass	Eq. (1.5)
m_l	mass of a limb	Eq. (1.12)
m_p	mass of the pectoral muscles	Eq. (1.24)
m_s	mass of the supracoracoideus muscles	Eq. (1.25)
$\tilde{n}$	amplification factor	Eq. (2.12)
N	threshold value that triggers turbulent flow in e^N method	Eq. (2.17)
p	static pressure	Eq. (2.5)
$\bar{p}$	normalized static pressure	Eq. (4.19)
P_{aero}	total aerodynamic power	Eq. (1.28)
p_{center}	pressure at the center of a vortex core rotating as a rigid body	Eq. (2.23)
P_{ind}	induced power (required for generating lift and thrust)	Eq. (1.30)
P_{iner}	inertial power (required for moving the wings)	Eq. (1.31)
P_{pro}	profile power (required for overcoming form and friction drag of the wings)	Eq. (1.30)
P_{par}	parasite power (required for overcoming form and friction drag of the body)	Eq. (1.30)
P_{tot}	total power required for flight	Eq. (1.31)
q_∞	far-field dynamic pressure	Eq. (3.12)
r_1	radius of the vortex core rotating as a rigid body	Eq. (2.23)
R	wing length	Eq. (4.20)
Re	Reynolds number	
Re_{f2}	Reynolds number for 2D flapping airfoils	Eq. (4.7)
Re_{f3}	Reynolds number for 3D flapping wing	Eq. (4.8)
Re_T	turbulent Reynolds number	Eq. (2.10)
Re_θ	momentum-thickness Reynolds number	Eq. (2.12)
Re_{θ_0}	critical Reynolds number	Eq. (2.12)
$Re_{\theta T}$	momentum-thickness Reynolds number at transition point	Eq. (2.19)
S	wing area	Eq. (1.3)
S	second Piola–Kirchoff stress tensor	Eq. (3.23)
S_0	membrane prestress	Eq. (3.7)
St	Strouhal number	Eq. (4.9)
T	wing-stroke time scale	Eq. (1.14)
T	thrust (for hovering)	Eq. (1.29)
T_i	free-stream turbulence intensity	Eq. (2.17)
t	time	Eq. (2.5)
U	forward-flight velocity (free-stream velocity)	Eq. (1.2)
U_{ref}	reference velocity	Eq. (1.1)
u_e	edge velocity	Eq. (2.2)
u_i	velocity vector in Cartesian coordinates	Eq. (2.4)

$\overline{u_i}$	normalized velocity vector in Cartesian coordinates	Eq. (4.19)
U_f	flapping velocity	Eq. (1.2)
U_{mp}	velocity for minimum power (forward flight)	Eq. (1.33)
U_{Mr}	velocity for maximum range (forward flight)	Eq. (1.33)
U_r	relative flow velocity	Eq. (1.2)
w	vertical velocity in the far wake	Eq. (4.22)
w_i	downwash (induced) velocity	Eq. (1.2)
W	weight	Eq. (1.3)
W	out-of-plane membrane displacement	Eq. (3.20)
W/S	wing loading	Eq. (1.7)
x_i	spatial coordinate vector	Eq. (2.4)
x_l	leg length	Eq. (1.19)
x_T	transition onset position	Eq. (2.19)
α	angle of attack	Eq. (3.1)
$\alpha(t)$	feathering angle (pitch angle) of a flapping wing	Eq. (4.3)
α_0	initial pitch angle at the beginning of the stroke	Eq. (4.5)
α_a	pitch amplitude	Eq. (4.5)
β	stroke-plane angle	Eq. (4.21)
γ	membrane tension	Eq. (3.3)
$\hat{\gamma}$	dimensionless membrane tension	Eq. (3.7)
Γ	circulation	Eq. (2.23)
δ^*	boundary-layer displacement thickness	Eq. (2.3)
$\overline{\delta}$	nominal membrane strain	Eq. (3.8)
ε	dimensionless excess length of a membrane	Eq. (3.2)
θ	boundary-layer momentum thickness	Eq. (2.1)
$\theta(t)$	elevation angle of a flapping wing	Eq. (4.2)
ζ, η	curvilinear coordinates along the membrane airfoil	Eq. (3.3)
ν	kinematic viscosity	Eq. (2.5)
ν_{Te}	effective eddy viscosity	Eq. (2.18)
ν_T	turbulent eddy viscosity	Eq. (2.6)
Π_1	aeroelastic parameter (elastic-strain-dominated membrane tension)	Eq.(3.16)
Π_2	aeroelastic parameter (pretension-dominated membrane tension)	Eq. (3.18)
$\phi(t)$	positional angle of a flapping wing	Eq. (4.1)
Φ	stroke angular amplitude	Eq. (4.7)
φ	phase difference between plunging and pitching motion	Eq. (4.4)
ρ	(air) density	Eq. (1.3)
τ_{ij}	Reynolds stress tensor	Eq. (2.6)
τ	tangential surface traction for 2D membrane	Eq. (3.4)
ω	angular velocity of a flapping wing $= 2\pi f$	Eq. (1.1)
ω	dissipation rate for k–ω turbulence model	Eq. (2.7)
ω	frequency	Eq. (2.21)
$\dot{\omega}$	angular acceleration	Eq. (1.13)

List of Abbreviations

Abbreviation	Definition
2D	two-dimensional
3D	three-dimensional
AoA	angle of attack
AR	aspect ratio
CFD	computational fluid dynamics
CSD	computational structural dynamics
DNS	direct numerical simulation
DPIV	digital particle-image velocimetry
LES	large-eddy simulation
LEV	leading-edge vortex
LSB	laminar separation bubble
MAV	micro air vehicle
RANS	Reynolds-averaged Navier–Stokes
TEV	trailing-edge vortex
TS	Tollmien–Schlichting
WTV	wingtip vortex

Preface

Low Reynolds number aerodynamics is important to a number of natural and manmade flyers. Birds, bats, and insects have been of interest to biologists for years, and active study in the aerospace engineering community has been increasing rapidly. Part of the reason is the advent of micro air vehicles (MAVs). With a maximal dimension of 15 cm and nominal flight speeds of around 10 m/s, MAVs are capable of performing missions such as environmental monitoring, survelliance, and assessment in hostile environments. In contrast to civilian transport and many military flight vehicles, these small flyers operate in the low Reynolds number regime of 10^5 or lower. It is well established that the aerodynamic characteristics, such as the lift-to-drag ratio of a flight vehicle, change considerably between the low and high Reynolds number regimes. In particular, flow separation and laminar–turbulent transition can result in substantial change in effective airfoil shape and reduce aerodynamic performance. Because these flyers are lightweight and operate at low speeds, they are sensitive to wind gusts. Furthermore, their wing structures are flexible and tend to deform during flight. Consequently, the aero/fluid and structural dynamics of these flyers are closely linked to each other, making the entire flight vehicle difficult to analyze.

The primary focus of this book is on the aerodynamics associated with fixed and flapping wings. Chapter 1 offers a general introduction to low Reynolds number flight vehicles, considering both biological flyers and MAVs, followed by a summary of the scaling laws, which relate the aerodynamics and flight characteristics to a flyer's size on the basis of simple geometric and dynamics analyses. In Chapter 2, we discuss the aerodynamics of fixed, rigid wings. Both two- and three-dimensional airfoils with typically low-aspect-ratio wings are considered. Chapter 3 examines structural flexibility within the context of fixed-wing aerodynamics. The implications of laminar–turbulent transition, multiple time scales, airfoil shapes, angles of attack, stall margin, and the structural flexibility and time-dependent fluid and structural dynamics are highlighted.

Unsteady flapping-wing aerodynamics is presented in Chapter 4, in particular, the interplay between flapping kinematics and key dimensionless parameters such as the Reynolds number, Strouhal number, and reduced frequency. The various unsteady lift-enhancement mechanisms are also addressed, including leading-edge vortex, rapid pitch-up, wake capture, and clap-and-fling.

The materials presented in this book are based on our own research, existing literature, and communications with colleagues. At different stages, we have benefited from collaborations and interactions with Peter Ifju, David Jenkins, Rick Lind, Raphael Haftka, Richard Fearn, Roberto Albertani, and Bruce Carroll of the University of Florida; Luis Bernal, Carlos Cesnik, and Peretz Friedmann of the University of Michigan; Michael Ol, Miguel Visbal, and Gregg Abate, and Johnny Evers of the Air Force Research Laboratory; Ismet Gursul of the University of Bath; Charles Ellington of Cambridge University; Keiji Kawachi of the University of Tokyo; Hikaru Aono of Chiba University; Max Platzer of Naval Postgraduate School; and Mao Sun of the Beijing University of Aeronautics and Astronautics. In particular, we have followed the flight vehicle development efforts of Peter Ifju and his group and enjoyed the synergy between us.

MAVs and biological flight is now an active and well-integrated research area, attracting participation from a wide range of talents. The complementary perspectives of researchers with different training and background enable us to develop new biological insight, mathematical models, physical interpretation, experimental techniques, and design concepts.

Thinking back to the time we started our own endeavor a little more than 10 years ago, we see that substantial progress has taken place, and there is every expectation that significantly more will advance in the foreseeable future. We look forward to it!

Wei Shyy, Yongsheng Lian, Jian Tang, Dragos Viieru
Ann Arbor, Michigan, U.S.A.

Hao Liu
Chiba, Japan

December 31, 2006

CHAPTER ONE

Introduction

Bird, bat, and insect flight has fascinated humans for many centuries. As enthusiastically observed by Dial (1994), most species of animals fly. There are nearly a million species of flying insects, and of the living 13,000 warm-blooded vertebrate species (i.e., birds and mammals), 10,000 (9000 birds and 1000 bats) have taken to the skies. With respect to maneuvering a body efficiently through space, birds represent one of nature's finest locomotion experiments. Although aeronautical technology has advanced rapidly over the past 100 years, nature's flying machines, which have evolved over 150 million years, are still impressive. Considering that humans move at top speeds of 3–4 body lengths per second, a race horse runs approximately 7 body lengths per second, a cheetah accomplishes 18 body lengths per second (Norberg, 1990), a supersonic aircraft such as the SR-71, "Blackbird," traveling near Mach 3 (~2000 mph) covers about 32 body lengths per second, it is amazing that a common pigeon (*Columba livia*) frequently attains speeds of 50 mph, which converts to 75 body lengths per second. A European starling (*Sturnus vulgaris*) is capable of flying at 120 body lengths per second, and various species of swifts are even more impressive, over 140 body lengths per second. The roll rate of highly aerobatic aircraft (e.g., the A-4 Skyhawk) is approximately 720°/s, and a Barn Swallow (*Hirundo rustics*) has a roll rate in excess of 5000°/s. The maximum positive G-forces permitted in most general aviation aircraft is 4–5 G and select military aircraft withstand 8–10 G. However, many birds routinely experience positive G-forces in excess of 10 G and up to 14 G. The primary reasons for such superior maneuvering and flight characteristics include the "scaling laws" with respect to a vehicle's size, as well as intuitive but highly developed sensing, navigation, and control capabilities. As McMasters and Henderson put it, humans fly commercially or recreationally, but animals fly professionally (McMasters and Henderson, 1980). Figure 1.1 illustrates several maneuvering characteristics of biological flyers; these capabilities are difficult to mimic by manmade machines. Combining flapping patterns, body contour, and tail adjustment, natural flyers can track target precisely and instantaneously. Figure 1.2 shows hummingbirds conducting highly difficult and precise flight control. To take off, natural flyers synchronize wings, body, legs, and tail. As shown in Figure 1.3, they can take off on water, from land, and off a tree, exhibiting varied and sophisticated patterns. While gliding, as shown in Figure 1.4, they flex their wings to control their speed as well as the direction. On landing, as depicted in Figure 1.5, birds fold their wings to reduce lift, and flap to accommodate wind gusts and to adjust for the location of the available landing area.

1

Figure 1.1. Maneuvering capabilities of natural flyers: (a) Canadian geese's response to wind gust; (b) speed control and target tracking of a seagull; (c) precision touchdown of a finch; (d) a hummingbird defending itself against a bee.
This figure is available in color for download from www.cambridge.org/9780521204019

Figure 1.2. Natural flyers can track target precisely and instantaneously. Shown here are hummingbirds using flapping wings, contoured body, and tail adjustment to conduct flight control.
This figure is available in color for download from www.cambridge.org/9780521204019

Figure 1.3. Natural flyers synchronize wings, body, legs, and tail to take off (top) on water, (middle) from land, and (bottom) off a tree.
This figure is available in color for download from www.cambridge.org/9780521204019

Since the late 1990s, the so-called micro air vehicles (MAVs) have attracted substantial and growing interest in the engineering and science communities. The MAV was originally defined as a vehicle with a maximal dimension of 15 cm or less, which is comparable to the size of small birds or bats, and a flight speed of 10–20 m/s (McMichael and Francis, 1997). Equipped with a video camera or a sensor, these

Figure 1.4. Birds such as seagulls glide while flexing their wings to adjust their speed as well as to control their direction.
This figure is available in color for download from www.cambridge.org/9780521204019

Figure 1.5. On landing, birds fold their wings to reduce lift, and flap to accommodate wind gusts and to adjust for their available landing area.
This figure is available in color for download from www.cambridge.org/9780521204019

vehicles can perform surveillance and reconnaissance, targeting, and biochemical sensing at remote or otherwise hazardous locations. With the rapid progress made in structural and material technologies, miniaturization of power plants, communication, visualization, and control devices, numerous groups have developed successful MAVs. Overall, alternative MAV concepts, based on fixed wing, rotary wing, and flapping wing, have been investigated. Figure 1.6(a) shows a 15-cm MAV designed by Ifju et al. (2002), which uses a fixed, flexible-wing concept. Figure 1.6(b) shows a rotary-wing MAV with 8.5-cm rotary diameter designed by Muren (http://www.proxflyer.com). Figure 1.6(c) shows a biplane MAV designed by Jones and Platzer (2006), which uses a hybrid flapping–fixed-wing-design, with the flapping wing generating thrust and the fixed wing producing necessary lift. Figure 1.6(d) shows a recent development by Kawamura et al. (2006) that relies on flapping wing to generate both lift and thrust and possesses some flight control capabilities.

Figure 1.7 highlights more detailed vehicle characteristics of flexible-wing MAVs designed by Ifju and coworkers. The annual International Micro Air Vehicle

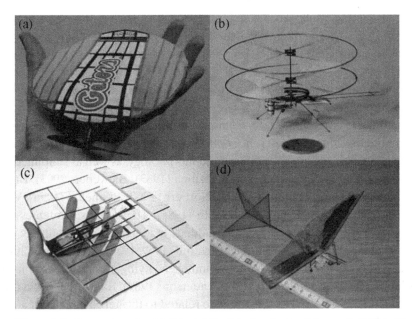

Figure 1.6. Representative MAVs: (a) flexible fixed wing (Ifju et al., 2002); (b) rotary wing (http://www.proxflyer.com); (c) hybrid flapping–fixed wing, with the fixed wing used for lift and the flapping wing for thrust (Jones and Platzer, 2006); and (d) flapping wing for both lift and thrust (Kawamura et al., 2006).
This figure is available in color for download from www.cambridge.org/9780521204019

Competition has offered a substantial forum, encouraging the development of MAVs. For example, one of the competition categories is to fly 600 m, capture an image of a 1.5 m × 1.5 m target, and transmit the image with telemetry. The smallest vehicle capable of successfully completing the mission is declared the winner. Since the first competition, the winning vehicle's size has drastically decreased, and now the maximum dimension is just barely over 10 cm.

The MAVs operate in the low Reynolds number regime (originally envisioned to be 10^4–10^5, now even lower), which, compared with large, manned flight

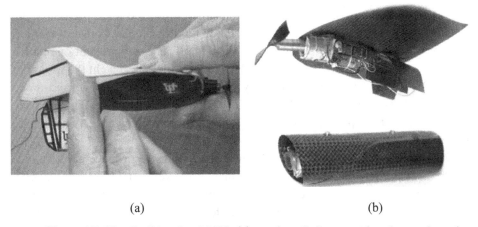

(a)   (b)

Figure 1.7. The flexible-wing MAVs (a) can benefit from passive shape adaptation in accordance with instantaneous aerodynamic loading, and (b) can be packed very easily based on need (courtesy Peter Ifju).

vehicles, have unfavorable aerodynamic characteristics, such as low lift-to-drag ratio (Lissaman, 1983). On the other hand, the MAVs' small geometric dimensions result in favorable scaling characteristics, such as reduced stall speed and better structural survivability.

There is great potential for collaborative research between biologists and engineers because MAVs and biological flyers share similar dimensions, weight, flight speeds, and flight environment. Substantial literature exists, especially in the biological community. General references offering broad accounts of biological flight, including geometric scaling laws, power, and morphology, as well as simplified modeling, can be found in Alexander (2002), Azuma (1983), Biewener (2003), Brodsky (1994), Dudley (2000), Grodnitsky (1999), Norberg (1990), Tennekes (1996), Videler et al. (2004), Vogel (1996), and Ward-Smith (1984). The symposia volumes edited by Wu et al. (1975), Pedley (1977), and Maddock et al. (1994) offer multiple angles related to flight as well as to swimming. Lighthill (1969, 1977), Wu (1971), Childress (1981), and Maxworthy (1979) discuss swimming and flying primarily from analytical viewpoints. Finally, the standard texts by Anderson (1989), Katz and Plotkin (2002), and Shevell (1983) present basic knowledge related to the aerodynamics of airplane flight. Our effort in this book is aimed at the aerodynamics relevant to both biological flyers and manmade MAVs.

In this chapter, we first introduce the flapping flight in nature, including the kinematics of flapping-wing vehicles and the lift- and thrust-generation mechanisms. Second, we present the scaling laws related to the mechanics and energetics of avian flight. Then we discuss drag and power related to avian flight. These two quantities are intimately connected. The different power components are presented separately and later summed together, giving the total power required for hovering and forward flight. A comparison between the power components for a fixed- and a flapping-wing vehicle is also presented. The results of these different power calculations are summarized in the form of power curves.

1.1   Flapping Flight in Nature

Flapping flight is more complicated than flight with fixed wings because of the structural movement and the resulting unsteady fluid dynamics. Conventional airplanes with fixed wings are, in comparison, very simple. The forward motion relative to the air causes the wings to produce lift. However, in biological flight the wings not only move forward relative to the air, they also flap up and down, plunge, and sweep (Dial, 1994; Goslow et al., 1990; Norberg, 1990; Shipman, 1998; Tobalske and Dial, 1996). Early photographs and some general observations are given by Aymar (1935) and Storer (1948).

While flapping, birds systematically twist their wings to produce aerodynamic effects in ways that the ailerons on the wings of conventional airplanes operate. Specifically, one wing is twisted downward (pronated), thus reducing the angle of attack (AoA) and corresponding lift, while the other wing is twisted upward (supinated) to increase lift. With different degrees of twisting between wings, a bird

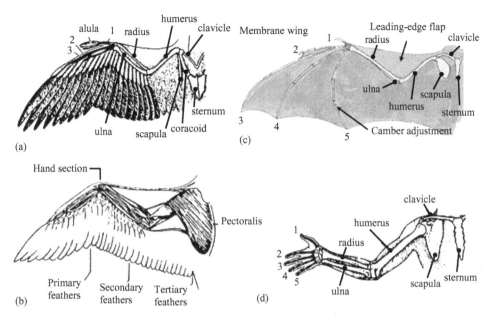

Figure 1.8. Schematics of (a), (b) a bird wing, (c) a bat wing, and (d) a human arm. For birds, the upper arm, the "humerus," is proportionately shorter, the "wrist" and "palm" bones are fused together for greater strength in supporting the primary flight feathers. For bats, the bone–membrane combination creates a leading-edge flap and allows passive camber adaptation in the membrane area. (a), (b), and (d) are modified from Dhawan (1991); (c) is adopted from Anders (2000).

is able to roll (Dial, 1994). For a bird to be able to deform and twist its wings, an adaptation in the skeletal and muscular systems is required. The key features that seem desirable are modification of camber and flexing of the wing planform between upstroke and downstroke, twisting, area expansion and contraction, and transverse bending. To perform these functions, birds have a bone structure in their wings similar to the one in a human arm. However, birds have more stringent muscle and bone movement during flight. Figure 1.8 shows a schematic of a bird wing compared with a human arm and hand. Figure 1.9 compares the cross-sectional shapes of a pigeon wing and a conventional transport airplane wing. The pigeon wing exhibits noticeably more variations in camber and thickness along the spanwise direction.

1.1.1 *Unpowered Flight: Gliding and Soaring*

Flying animals usually flap their wings to generate both lift and thrust. But if they stop flapping and keep their wings stretched out, their wings actively produce only lift, not thrust. Thrust can be produced by gravity force while the animal is descending. When this happens, we call them gliders. In addition to bats and larger birds, gliders can also be found among fish, amphibians, reptiles, and mammals.

To maintain level flight, a flying animal must produce both lift and thrust to balance the gravity force in the vertical direction and drag in the horizontal

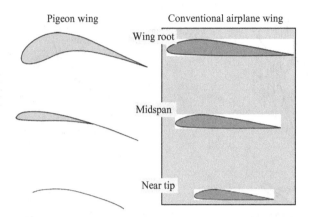

Figure 1.9. Comparison of cross-sectional shapes of a pigeon wing and a conventional transport airplane wing. The pigeon wing exhibits noticeably more variations in camber and thickness along the spanwise direction.

direction, respectively. Because gliding occurs with no active thrust production, an animal always resorts to the gravity force to overcome the drag. In gliding, the animal tilts its direction of motion slightly downward relative to the air that it moves through. When the animal tilts downward, the resulting angle between the motion direction and the air becomes the gliding angle. The gliding angle directly controls the lift-to-drag ratio. The higher this ratio, the shallower the glide becomes. Recall from basic fluid dynamics that the lift-to-drag ratio increases with the Reynolds number, a parameter proportional to animal size and flight speed. Large flying animals fly at high Reynolds numbers and have a large lift-to-drag ratio. For example, a wandering albatross, with a wing span of over 3 m, has a reported lift-to-drag ratio of 19 whereas the fruit fly, which has a span of 6 mm, has a ratio of 1.8 (Alexander, 2002). If the animal has a low lift-to-drag ratio, it must glide (if it can) with a considerably large glide angle. For example, a lizard in the Southeast Asian genus *Draco* has a lift-to-drag ratio of 1.7 and it glides at an angle of 30°; a North American flying squirrel has a glide angle of about 18°– 26° with a lift-to-drag ratio of 2 or 3 (Alexander, 2002).

While gliding animals take a downward tilt to have the gravity-powered flight, many birds can ascend without flapping their wings, and this is called soaring. Instead of using gravity, soaring uses energy in the atmosphere, such as rising air currents (Alexander, 2002).

1.1.2 *Powered Flight: Flapping*

An alternative method to gliding used by many biological flyers to produce lift is flapping-wing flight. The similarities between the aerodynamics of a flapping wing and that of a rotorcraft, although limited, can illustrate a few key ideas. Take for example the rotors of a helicopter, which rotate about the central shaft continuously. The relative flow around the rotors produces lift. Likewise, a flapping wing

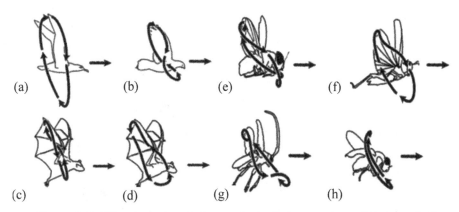

Figure 1.10. Wingtip paths relative to the body for a variety of flyers, as indicated by the arrows: (a) albatross, fast gait; (b) pigeon, slow gait; (c) horseshoe bat, fast flight; (d) horseshoe bat, slow gait; (e) blowfly; (f) locust; (g) June beetle; (h) fruit fly. Adopted from Alexander (2002).

rotates, swings in an arc around its shoulder joint, and reverses direction every half-stroke. Helicopters and biological flyers use similar techniques to accelerate from hovering to forward flight as well. Helicopters tilt the rotational plane of rotors from horizontal to forward. The steeper the tilt of the rotor, the faster the helicopters accelerate. Biological flyers also tilt their flapping stroke plane: down and forward on the downstroke, and up and backward on the upstroke. To fly faster, biological flyers make the stroke more vertical by increasing the up-and-down amptitude of the movements. When biological flyers decrease their speed, they tend to flap their wings more horizontally, similar to the way helicopters change their rotors.

Birds, bats, and insects apply a variety of different flapping patterns in hovering and forward flight to generate lift and thrust. Larger birds have relatively simple wingtip paths. For example, an oval tip path is often associated with albatrosses (see Figure 1.10). Smaller flyers exhibit more complicated flapping patterns. Figure 1.10 illustrates the highly curved tip paths of a locust and a fruit fly, the figure-eight pattern of a pigeon, and the more complicated paths of June beetles and blowflies.

1.1.3 *Hovering*

Whether a flying animal can hover or not depends on its size, moment of inertia of the wings, degrees of freedom in the movement of the wings, and the wing shape. As a result of these limitations, hovering is mainly performed by smaller birds and insects. Larger birds can hover only briefly. Although some larger birds like kestrels seem to hover more regularly, in fact, they use the incoming wind to generate enough lift. There are two kinds of hovering, symmetric hovering and asymmetric hovering, as described by Weis-Fogh (1973) and Norberg (1990).

Figure 1.11. Selected seagull wing configurations during flapping, which show various stages of strokes. Note that the wings are often flexed with their primaries rotated.

For larger birds, which cannot rotate their wings between forward and backward strokes, the wings are extended to provide more lift during downstroke, whereas during the upstroke the wings are flexed backward to reduce drag. In general the flex is more pronounced in the slow forward flight than in fast forward flight. This type of asymmetric hovering is usually called "avian stroke" (Azuma, 1992) and is illustrated in Figure 1.11. As shown in the figure, to avoid large drag forces and negative lift forces, these birds flex their wings during the upstroke by rotating the primaries (tip feathers) to let air through.

Symmetric hovering, also called normal or true hovering, or "insect stroke," is performed by hummingbirds or insects that hover with fully extended wings during the entire wing-beat cycle. Lift is produced during the entire wing stroke, except at the reversal points. The wings are rotated and twisted during the backstroke so that the leading edge of the wing remains the same throughout the cycle, but the upper surface of the wing during the forward stroke becomes the lower surface during the backward stroke. The wing movements during downstroke and upstroke can be seen in Figure 1.12. Note that, during hovering, the body axis is inclined at a desirable angle and the wing movements describe a figure of a lying eight in the vertical plane.

1.1.4 *Forward Flight*

When a natural flyer's aerodynamic performance is analyzed, an important parameter is the ratio between the forward velocity and the flapping velocity, which is expressed in terms of the reduced frequency:

$$k = \frac{\omega c}{2U_{\text{ref}}}, \tag{1.1}$$

where ω, c, and U_{ref} are, respectively, the angular velocity of a flapping wing, the wing's reference chord, and the reference velocity, in this case the flyer's forward-flight velocity. The unsteady effects increase with increasing reduced frequency, and

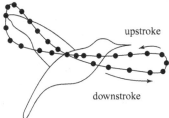

Figure 1.12. Illustration of biological flapping-wing patterns: forward and back strokes, and flexible- and asymmetric-wing motions (photos by the authors), and figure-eight pattern (Azuma 1992).
This figure is available in color for download from www.cambridge.org/9780521204019

therefore, depending on the forward velocity, different techniques have been devised to calculate the forces acting on a specific species.

In slow forward flight, both reduced frequency and wing-beat amplitude tend to be high, resulting in highly unsteady flow structures. In accordance with the Lifting Line Theory (Jones, 1990), the lift on a wing is related to the strength of the bound vortex. The trailing vortices (the tip vortices) are of the same circulation magnitude as the bound vortex. At the beginning/end of the downstroke, when the flapping velocity changes direction, a transverse vortex (starting/stopping vortex) is produced at the trailing edge, and, according to Kelvin's circulation theorem, these two transverse vortices connect the two tip vortices and result in the shedding of a vortex ring. Some flyers (for example, doves) make use of the clap-and-fling mechanism to generate the starting vortex and reduce the delay in building up maximum lift during the first part of the downstroke.

In fast forward flight, the reduced frequency and the wing-beat amplitude tend to be low, and the wake consists of a pair of continuous undulating vortex tubes or

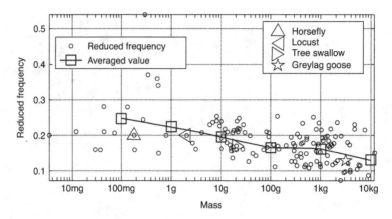

Figure 1.13. Mass versus reduced frequency for birds and insects.

line vortices approximately behind the wingtips. In such cases, it is not unusual that the outer part of the wing is folded to align with the free-stream direction (to reduce drag), and only the arm-wing contributes to aerodynamic lift during the upstroke.

When flying animals' lift and thrust are evaluated, depending on the magnitude of the reduced frequency, either unsteady or quasi-steady methods can be used. Early work by Ellington (1984a) has shown that quasi-steady analysis substantially underpredicts the aerodynamic force needed to sustain the insect weight. As will be discussed in Chapter 4, much of recent flapping-wing research has focused on the understanding of unsteady aerodynamic mechanisms resulting from wing movement. Figure 1.13 shows the correlation between a flyer's mass and the reduced frequency. The data are based on those reported by Azuma (1992) and Pennycuick (1989), aided by the cruising-velocity estimate documented by Tennekes (1996). Overall, the reduced frequency decreases as the size and mass grow, indicating that small flyers use more unsteadiness in their flight than large flyers. Although this figure does not explain how the unsteadiness is used, it does disclose that unsteadiness plays a critical role in small flyers' movement.

To quantify the lift and thrust generated by the flapping motion requires more sophisticated tools. However, we can understand the role of the unsteady effects by examining the relationship between the forward velocity and the flapping velocity in terms of the reduced frequency. It is also noted that different sections of wing function differently in force generation. We can better understand this concept by introducing the relative flow velocity U_r, defined in vector notation as,

$$U_r = U + U_f + w_i. \tag{1.2}$$

Here, U is the forward velocity of the flyer, U_f is the flapping velocity, and w_i is the downwash (induced) velocity. The relative velocity determines the aerodynamic forces on the wing. For fast forward flight the downwash velocity is small and can be largely neglected. With a larger wing span, U_f increases and changes its direction, which affects the magnitude and direction of U_r. Because U_r determines the resulting

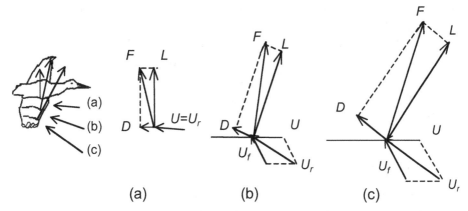

Figure 1.14. Velocity-force vector diagram at different flapping wingspan locations for fast forward flight. Here, the lift and drag are defined based on the effective velocity combining forward and local flapping velocities. For the entire vehicle, the lift is defined to be normal to the forward velocity (U), i.e., in the vertical direction, and drag/thrust in the horizontal direction. According to the resulting force vector F illustrated here, drag of the vehicle is generated by the inner wing, and thrust of the vehicle is generated by the outer wing.

aerodynamic force F acting on each wing section along the span, F will also change in magnitude and direction. The changes can be seen in Figure 1.14.

It is commonly held that during the downstroke the inner part of the wing produces lift and drag, whereas the outer part produces lift and thrust. The net aerodynamic force produced by the wings during a downstroke is directed upward and forward, providing both lift and thrust. To obtain this favorable force distribution, the wings have to be twisted. When the wings are twisted, an optimal relative velocity can be obtained at each wing section, throughout the wing stroke. Because the relative velocity determines the direction of the resultant aerodynamic force, this force will be directed backward at the wing root and gradually turned forward when moving along the wingspan. At the wingtip region the resultant aerodynamic force points toward the forward direction, giving both lift and thrust. However, such a picture is imprecise and needs to be examined carefully on a case by case basis.

Flying animals use different mechanisms for various missions such as take-off, landing, or gliding. Even for forward flight, they change their wing and body movements while flying through a range of speeds. Tobalske and Dial (1996) analyzed videotapes of black-billed magpies (*Pica pica*) flying at speeds of 4–14 m/s and pigeons (*Columbia livia*) flying at 6–20 m/s in a wind tunnel. Pigeons have higher wing loading and higher aspect ratio wings compared with magpies. Both species alternate phases of steady-speed flight with phases of acceleration and deceleration, particularly at intermediate flight speeds. The birds modulate their wingbeat kinematics among these phases and frequently exhibit nonflapping phases while decelerating. They find that, during steady-speed flight, wing-beat frequency does not change appreciably with horizontal flight speed. Instead, the body angle relative to the horizontal decreases with increasing flight speed, thereby illustrating that the dominant function of wing flapping changes from weight support at slow speeds to positive thrust at fast speeds. Pigeons progressively flex their wings during glides as flight speed increases but never perform bounding. Wingspan during glides in magpies does not vary with flight speed, but the percentage of bounding among nonflapping intervals increases with speed from 10 to 14 m/s. The use of nonflapping wing postures seems to be related to the gaits used during flapping and to the AR of the wings.

1.2   Scaling

When studying natural flyers, it is insightful to assess the effects of different parameters, such as wing area and wingspan, on the flight characteristics, based on dimensional analysis (Lighthill, 1977; Norberg, 1990; Pennycuick, 1992; Schmidt-Nielsen, 1984). Tennekes (1996) presents very interesting correlations to summarize the various scaling laws ranging from birds and insects to aircraft. He considered the relations among cruising speed, weight, and wing loading, and established *The Great Flight Diagram*. The diagram is shown in Figure 1.15.

With technical advancement, the MAV dimensions, wing loading, and speed will continue to decrease, moving toward the lower left-hand corner in Figure 1.15. From Figure 1.15 one can compare and correlate relations between species with a pronounced difference in size. For example, the small fruit fly *Drosophila melanogaster* can be compared to the Boeing 747, which weighs about 500 billion times more. By using scaling analysis, one can predict how a parameter such as wingspan varies with another parameter such as the body mass for natural flyers in general or specific animal groups.

As an illustration, consider the balance between lift and weight during steady-state flight,

$$L = W = 1/2\rho U^2 S C_L. \tag{1.3}$$

From Eq. (1.3) it is possible to get an understanding of how wing area, airspeed, density, and wing loading are connected.

Wing area (S): The wing area for a flight vehicle is often defined as the area projected when the wing is seen from above and usually the area includes the contribution from the "wing area" inside the fuselage.

Air speed (U): The air speed is defined as the forward velocity for the flight vehicle. Given a particular AoA, a twofold increase in speed will result in a fourfold increase in lift.

Density (ρ): For cases of interest to bird flight, the density of the air is basically unchanged because birds fly within a narrow altitude near sea level. In general, a decrease of density that is due to an increase in altitude will decrease the lift.

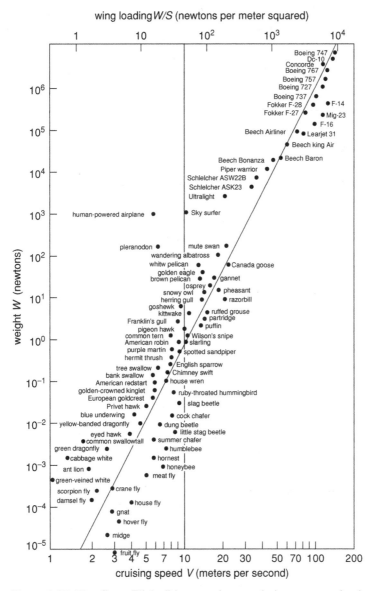

Figure 1.15. The Great Flight Diagram gives a relation among wing loading, weight, and cruising speed. Adopted from Tennekes (1996).

Wing loading (*W/S*): From Eq. (1.3), it is clear that the cruising speed depends on the wing loading:

$$\frac{W}{S} = \frac{\rho\, U^2}{2} C_L. \tag{1.4}$$

Equation (1.4) shows that, the greater a flyer's wing loading, the faster it has to fly. Some of the relations among body mass and parameters connected to birds are shown in Table 1.1. Figure 1.15 offers a correlation between sizes and speeds, and

Table 1.1. *Power functions of wing dimensions and flight parameters against body mass m. Originally compiled by Greenewalt (1975), Norberg (1990), and Rayner (1979b, 1988).*

Animal groups	Dimensions	Correlation (based on geometrical similarities)	All birds (based on empirical data)	All birds except hummingbirds (based on empirical data)	Hummingbirds (based on empirical data)
Wing span	m	$m^{0.33}$	–	$1.17m^{0.39}$	$2.24m^{0.53}$
Wing area	m²	$m^{0.67}$	–	$0.16m^{0.72}$	$0.69m^{1.04}$
Wing loading	N/m²	$m^{0.33}$	–	$62.2m^{0.28}$	$17.3m^{-0.04}$
Aspect ratio	–	0	–	$8.56m^{0.06}$	$7.28m^{0.02}$
Minimum power speed U_{mp}	m/s	$m^{0.17}$	$5.70m^{0.16}$	–	–
Minimum power P_{mp}	W	$m^{0.17}$	$10.9m^{0.19}$	–	–
Minimum cost of transport C_{min}	–	0	$0.21m^{-0.07}$	–	–
Wing-beat frequency f_w	Hz	$m^{-0.33}$	$3.87m^{-0.33}$	$3.98m^{-0.27}$	$1.32m^{-0.60}$

Table 1.1 summarizes expanded correlations. More details will be discussed in the following sections.

1.2.1 Geometric Similarity

The concept of geometric similarity can help relate different physical quantities by means of the dimensional argument. Under the assumption of geometric similarity, Figure 1.16 correlates wing loading with the weight of a vehicle. For example, the wing loading is proportional to one-third power of the weight, if the aerodynamic parameters remain unchanged (which is not true, as will be discussed in detail). If flyers are assumed to be geometrically similar, the weight W, lift L, and mass m for unaccelerated level flight, can be expressed with respect to a characteristic length l as

$$W = L = mg. \tag{1.5}$$

The wing area S and bird weight are expressed as

$$S \sim l^2, \quad W \sim l^3. \tag{1.6}$$

Then the wing loading can also be expressed as

$$\frac{W}{S} = k_1 W^{1/3}, \tag{1.7}$$

where k_1 is a constant to be determined empirically. Liu (2006) shows that a suitable value of k_1 is 53 and 30.6 for aircraft and birds, respectively. The correlation is shown in Figure 1.16.

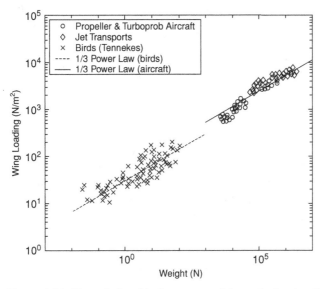

Figure 1.16. The relationship between weight and wing loading. Adopted from Liu (2006).

1.2.2 *Wingspan*

Often, when flapping animals are studied, parameters of interest are related to the body mass m of the animal. Using the dimensional argument method, assuming geometric similarity for the animals considered, one can determine a relation between the wingspan and the mass. Collecting data from Tennekes (1996) for birds ranging from a 0.026-N black-chinned hummingbird to a 116-N mute swan, and data for propeller/turboprop aircraft and jet transports published by Jackson (2001) covering a broad spectrum of aircraft from a 1500-N ultralight to an 1800-kN Boeing 747-400, Liu (2006) suggests that, over a large range of the weight, birds and aircraft basically follow the power law:

$$l = 1.654m^{1/3}(\text{aircraft}); \quad l = 1.704m^{1/3}(\text{birds}). \tag{1.8}$$

1.2.3 *Wing Area*

Norberg (1990) reports that the wing area between groups of animals shows larger variations than the wingspan. The departure from the geometrical relation is obvious, which is shown in Table 1.1. As for the wingspan, hummingbirds have the largest deviation from the geometrical relation. Hummingbirds seem to have a larger wing area for a given body mass compared with that of birds in general. From the variation of the wing area for different groups of birds, Greenewalt (1975) subdivided birds into different classes or "models." His model offers the following correlations:

1. The Passeriform model, (herons, falcons, hawks, eagles, owls): $S \sim m^{0.78}$.
2. The Shorebird model, (doves, parrots, geese, swans, albatross): $S \sim m^{0.71}$.
3. The Duck model, (grebes, loons, coots): $S \sim m^{0.78}$.

Table 1.2. *Weight, wing area, wing loading, and airspeeds for various seabirds, which are assumed to be geometrically similar. Data originally compiled by Tennekes (1996).*

Seabird	Weight W (N)	Wing area S (m^2)	Wing loading W/S	Air speed m/s	Air speed mph
Common tern	1.15	0.05	23	7.8	18
Dovs prion	1.7	0.046	37	9.9	22
Black-headed gull	2.3	0.075	31	9	20
Black skimmer	3	0.088	34	9.4	21
Common gull	3.67	0.115	32	9.2	21
Kittiwake	3.9	0.101	39	10.1	23
Royal tern	4.7	0.108	44	10.7	24
Fulmar	8.2	0.124	66	13.2	30
Herring gull	9.4	0.181	52	11.7	26
Great skua	13.5	0.214	63	12.9	29
Great black-backed gull	19.2	0.272	71	13.6	31
Sooty albatross	28	0.34	82	14.7	33
Black-browed albatross	38	0.36	106	16.7	38
Wandering albatross	87	0.62	140	19.2	43

These relations are consistent with those presented in Table 1.1 for all birds other than hummingbirds.

1.2.4 *Wing Loading*

Regarding wing loading, although the overall correlation shown in Eq. (1.7) seems reasonable, Greenewalt (1975) found that, in many cases, the relation between wing loading and mass increases slower than indicated in Eq. (1.7). For example, the three different families of birds, i.e., the Passeriforms, the Shorebirds, and Ducks, do not follow the 1/3 law. As indicated in Table 1.1, for hummingbirds, the wing loading is almost independent of body mass; hence different species can have the same wing loading. Tennekes (1996) utilized the data collected by Greenewalt (1975) and summarized the various scaling relations for seabirds, shown in Table 1.2. All gulls and their relatives have long, slender wings and streamlined bodies, so it was reasonable to assume geometric similarity. From Table 1.2 it is obvious that the wing loading and cruising speed generally increase with weight.

1.2.5 *Aspect Ratio*

As for aircraft, the aspect ratio (AR) can give indications of the flight characteristics for flapping animals. The AR is a relation between the wingspan b and the wing area S:

$$AR = \frac{b^2}{S}. \tag{1.9}$$

In general, the agility and ability to maneuver improves with a smaller AR. This is one of the reasons why military fighter aircraft and aerobatic airplanes have relatively short wingspans compared with those of conventional aircraft. The same relation is found for animals. Another consideration is that the induced drag, which is caused by the lift, tends to decrease with a higher AR. Obviously, the minimum induced drag is obtained with an infinitely long wing. Similarly, with a large AR, the lift-to-drag ratio *L/D*, the so-called glide ratio, increases with an increasing AR. The largest AR for birds is found among species that typically spend a substantial portion of their time in soaring flight instead of flapping. A typical example is the *wandering albatross (Diomeda exulans)*, which has an AR of about 15. According to Tennekes (1996) the glide ratio for the wandering albatross is around 20, compared to a modern sailplane with a glide ratio of around 60. Some readers might think the difference is big when comparing an albatross to a sailplane, but then it is worth noting that the sailplanes of today are often precisely made, and hence sensitive to airflow disturbances over the wing. Bugs attached to the wing surface might rapidly decrease the glide ratio by 30%. One should also note that the large raptors have long wingspans, but wide wings, and not particularly large AR.

1.2.6 *Wing-Beat Frequency*

The main function of wing bones is to transmit force to the external environment during flight. This force can, however, not be too high when there is a risk for bone or muscle failure. These limitations, along with the power available from flight muscles, settle the upper and lower limits of wing-beat frequency for flapping animals (Kirkpatrick, 1994; Pennycuick, 1989, 1996). Based on the insight into the flapping frequency, it is possible to estimate the power output from a bird's flight muscles and achieve an estimation of the power required for flying. According to Pennycuick (1975) it is possible to estimate the maximum flapping frequency $f_{w,\max}$ for geometrically similar animals, as shown in the following discussion. Because the force F_m exerted by a muscle is assumed to be proportional to the cross-sectional area of its attachment, we get

$$F_m \sim S \sim l^2. \tag{1.10}$$

Pennycuick (1975) assumes that the stresses in muscles and bones are constant and that the torque acting about the center of rotation of the proximal end of the limb, with length *l*, can be expressed as

$$J_T = F_m l. \tag{1.11}$$

The moment of inertia of the limb can now be determined as follows:

$$I = m_l \left(\frac{l}{2}\right)^2 \sim l^5. \tag{1.12}$$

The mass of the limb is denoted by m_l, and it is assumed that the limb has a uniform density.

The muscle in action has an angular acceleration, which can be characterized as

$$\dot{\omega} = \frac{J_T}{I} \sim \frac{l^3}{l^5} = l^{-2}. \tag{1.13}$$

From expression (1.13) it is easy to determine the stroke time scale T, and because the frequency f is proportional to $f \sim T^{-1}$ we get,

$$T = \dot{\omega}^{-1/2} \Rightarrow f \propto \dot{\omega}^{1/2}. \tag{1.14}$$

A relation between the body mass m and the maximum wing-beat frequency f_{max}, can also be derived:

$$f_{max} \sim T^{-1} \sim l^{-1} \sim m^{-1/3}. \tag{1.15}$$

With the assumption of geometric similarity, this is the upper limit of the flapping frequency. For the lower-flapping-frequency limit, which is the case for most birds in slow forward flight or hovering, the induced velocity w_i, which is the airflow speed in the wake right beneath the animal, dominates. Still, the weight W of the flyer must be balanced by the lift L and, referring to Eq. (1.3), we obtain the following relation for the induced velocity w_i:

$$W = L = \frac{\rho w_i^2 S C_L}{2} \Rightarrow w_i = \sqrt{\frac{2mg}{\rho C_L S}}. \tag{1.16}$$

The angular velocity of the wings can be dimensionally expressed as

$$\omega \sim w_i/l, \tag{1.17}$$

where l is a characteristic length, and with relations (1.5), (1.6) and (1.17), we obtain the final expression for the lower flapping limit as

$$f_{min} \sim \omega_{min} = \frac{w_i}{l} = \frac{1}{l}\sqrt{\frac{2mg}{\rho C_L S}} = \sqrt{\frac{2mg}{\rho C_L l^4}} \sim \left(\frac{l^3}{l^4}\right)^{1/2} = l^{-1/2} \sim m^{-1/6}. \tag{1.18}$$

Because of these two physical limits, animal flight has an upper and a lower bound for the flapping frequency.

1.3   Power Implication of a Flapping Wing

One of the first researchers to explore the consequences of the trend whereby larger animals oscillate their limbs at lower frequencies than smaller ones of similar type was Hill (1950). He concludes that the mechanical power produced by a particular flight muscle is directly proportional to the contraction frequency. This has made flapping frequency an important parameter when one is trying to describe the theories behind flapping wings. Pennycuick (1990) conducted one of the most thorough studies of the wing-beat frequency. He assumed that there is a *natural frequency* imposed on the animal by physical characteristics of its limbs and the forces that it must overcome. To be efficient, locomotion muscles have to be adapted to work at a particular frequency. For walking animals, Alexander (1976) showed that the natural frequency is proportional to

$$f \sim \sqrt{\frac{g}{x_l}},\tag{1.19}$$

where g is the acceleration of gravity and x_l is the leg length.

Pennycuick (1990) identified several physical variables that affect the wing-beat frequency:

b, wingspan (m),

S, wing area (m^2),

I, wing moment of inertia (kg m^2) $\sim mb^2$,

ρ, air density (kg/m^3).

Allowing the preceding variables to vary independently and assuming that the wing moment of inertia is proportional to mb^2, Pennycuick (1990) deduced the following correlation for the wing-beat frequency by considering 32 different species:

$$f = 1.08\,(m^{1/3}g^{1/2}b^{-1}S^{-1/4}\rho^{1/3}).\tag{1.20}$$

In an updated study Pennycuick (1996) added another 15 species and made a more detailed analysis, leading to the following expression:

$$f = m^{3/8}g^{1/2}b^{-23/24}S^{-1/3}\rho^{3/8}.\tag{1.21}$$

Equation (1.21) can be used to predict the wing-beat frequency of species whose mass, wingspan, and wing area are known. As mentioned earlier, the moment of inertia I for the wing is dependent on both the span b and the body mass m, and hence a change in any of these variables will result in a change in moment of inertia. This may not be appropriate, because a change in, for instance, body mass does not necessarily affect I. Therefore, if we intend to predict these effects on the wing-beat frequency, it is more suitable to include I as an independent parameter (Pennycuick, 1996):

$$f = (mg)^{1/2}\,b^{-17/24}S^{-1/3}I^{-1/8}\rho^{3/8}.\tag{1.22}$$

Another relation observed by Pennycuick et al. (1996) is the effect of air speed on wing-beat frequency when body mass changes. The data for the frequency and the airspeed U are fitted with a least-squares method:

$$f = k_2 + k_3/U + k_4U^3,\tag{1.23}$$

where k_2, k_3 and k_4 are proportional constants.

1.3.1 *Upper and Lower Limits*

Can scaling arguments provide any information about limits on the size of flapping flyers capable of sustained flight? As mentioned before, large pterosaurs once flew long ago. Some of these species were much larger than birds of today and there are discussions about whether they were able to flap or only to soar (Norberg, 1990). There are many parameters to consider when flapping flight is studied, but limitations to this kind of flight mainly depend on the power available and structural limits.

These limitations are intimately connected, as flapping frequency f affects both power and structural limits. To generate the power required for flight, most birds and other flapping animals have well-developed flight muscles. For birds these muscles are the pectoral muscles, powering the downstroke of the wings, and the supracoracoideus muscles powering the upstroke. Much effort has been made to determine power output and frequency levels, and to compare the masses of these muscles with the mass of the whole specimen. According to Rayner (1988) relations between body mass m and the mass of the pectoral and supracoracoideus muscles, m_p and m_s respectively, can be expressed as

$$m_p = 0.15m^{0.99}, \tag{1.24}$$

$$m_s = 0.016m^{1.01}. \tag{1.25}$$

This means that the flight muscles constitute approximately 17% of the total weight. The corresponding value for humans is 5%, according to Collins and Graham (1994). The power output from bird and "fast" human muscles is about the same, 150 W/kg. Because the wings are often flexed during the upstroke and therefore not exposed to the same aerodynamic force or moment of inertia as during the downstroke, the weight of the supracoracoideus muscle is generally low compared with the weight of the pectoral muscle. Hummingbirds are different, having an aerodynamically active upstroke (producing lift). In their case, the weight of the supracoracoideus is higher; according to Norberg (1990), this muscle group can constitute up to 12% of the body weight. The smallest supracoracoideus muscles are found among species with big spans, in which the muscle mass is about 6% of the total mass. This value is comparable to that of the human body, and hence these species have difficulties taking off without a headwind, running start, or a slope start from a height. However, species with long wings are usually able to soar, so the duration of the flapping-flight mode can therefore be decreased.

Pennycuick (1969, 1975, 1986) defined the power margin as the ratio of the power available from the flight muscles to that required for horizontal flight at the *minimum power speed*. As already mentioned, the power available depends on the flapping frequency that determines the upper and lower limits of the size of flying animals. Pennycuick (1986) concluded that the upper limit for flapping flight, based on actual sizes of the largest birds with powered flight, is a body mass of about 12–15 kg. Larger birds do not have the possibility of beating their wings fast enough to generate lift to sustain horizontal flight. Smaller birds have the advantage of being able to use different flapping frequencies, but for animals with a weight of about 1 g, there is another upper limitation. Their muscles need time to reset the contractile mechanism after each contraction (Norberg, 1990). For insects with wing-beat frequencies up to

400 Hz, this problem is solved with special fibrillar muscles capable of contracting and resetting at very high frequencies. This limitation results in a minimum mass for birds of 1.5 g and for bats, 1.9 g. The upper and lower wing-beat frequencies are also restricted because of the structural limits. Bones, tendons, and muscles are not capable of performing wing motions above a certain wing-beat frequency. Wing bones that have to transmit forces to the external environment during flight must be strong enough to not fail under the imposed loads. This means that the bones have to be stiff and strong and at the same time not too heavy. Kirkpatrick (1994) investigated the scaling relationships among body size and several morphological variables of bird and bat wings in order to estimated the stress levels in their wings. He also estimated the bending, shearing, and breaking stresses in the wing bones during flight. He suggested that the breaking stress for a bat humerus bone is around 75 MPa and for birds 125 MPa. This structural limit helps explain why no bat weighs more than 1.5 kg. Kirkpatrick (1994) found no relationship between either bending or shearing stresses and wingspan during gliding flight or during the downstroke in hovering. In general the safety factors are greater for birds than for bats. Hence birds are more capable of withstanding higher wing loading. A final conclusion by Kirkpatrick (1994) is that the stresses examined are scale independent.

1.3.2 *Drag and Power*

Like an aircraft, a natural flyer has to generate power to produce lift and to overcome drag during flight. When soaring or gliding without flapping, the flyer produces much of the power required by converting potential energy to kinetic energy, and vice versa. When flapping, the power is the rate at which work is produced by the flight muscles. For basic aerodynamic concepts discussed in this subsection, please refer to the standard textbooks such as those by Anderson (1989) and Shevell (1983).

The total aerodynamic drag (D_{aero}), acting on a flight vehicle, is a result of the resistance to the motion through the air. This total aerodynamic drag can be divided into different components. The two drag components acting on a wing in steady flight are the induced drag (D_{ind}), which is the drag that is due to lift, and the profile drag (D_{pro}), which is the drag associated with form and friction drag on the wing. The drag on a finite wing (D_w) is the sum of these two components:

$$D_w = D_{ind} + D_{pro}. \tag{1.26}$$

The parasite drag (D_{par}), which is defined as the drag on the body and *only* on the body, also contributes to the total drag on the bird. This drag component is due to the form and friction drag of the "nonlifting" body (it is true that, if the body is tilted at an angle to the free stream, it will contribute to lift, but this contribution is very small and is neglected). If the drags of the wing and body are summed, the total aerodynamic drag of the bird can be expressed as

$$D_{aero} = D_{ind} + D_{pro} + D_{par}. \tag{1.27}$$

The different powers subsequently presented are defined as the powers needed to overcome specific drags at a certain velocity. One obtains the total aerodynamic power required for steady forward flight by multiplying the drag force with the forward velocity (U_{ref}):

$$P_{aero} = D_{aero}U_{ref}. \tag{1.28}$$

The main effort here is to describe the different methods used for determining the total power required (P_{tot}) for flight. Depending on the forward velocity, the power components are calculated in different ways. There exists a clear difference between flight at zero velocity (hovering flight) and forward flight. Therefore we deal with these two cases separately when calculating the power components.

In hovering flight, the resulting velocity is essentially the same as the induced velocity (w_i) because of negligible forward velocity. In this case, the lift is equal to the thrust T, namely, the mass times the acceleration of gravity, and the total aerodynamic power required for hovering flight is

$$P_{aero} = Tw_i. \tag{1.29}$$

For forward flight, there exist three different power components corresponding to the three drag components in Eq. (1.27). The three components are the induced power (P_{ind}), which is the rate of work required for generating a vortex wake whose reaction generates lift and thrust, the profile power (P_{pro}), the rate of work needed to overcome form and friction drag of the wings, and the parasite power (P_{par}), which is the rate of work needed to overcome form and friction drag of the body. In the same way as for the drag components, the power components are added together to produce *the total aerodynamic power* (P_{aero}) required for horizontal flight:

$$P_{aero} = P_{ind} + P_{pro} + P_{par}, \tag{1.30}$$

where P_{ind} is the power needed for lift production during flight and decreases with increasing flight velocity. In the theory developed by Rayner (1979c), the upstroke is assumed not to contribute to any useful aerodynamic forces and is therefore not included. The wings are considered to move in only the stroke plane (i.e., no forward or backward movement). The induced power is calculated from the kinetic-energy increment in the wake from a single stroke. The shed vortex rings are elliptical and inclined at an angle to the horizontal. The kinetic energy has two components, the self-energy of the newly generated ring and the mutual energy of the new ring with each of the existing rings in the wake. The mutual energy contribution decreases with higher forward velocities and can be neglected for velocities above the minimum power required velocity. With this method the induced power can be calculated as a function of the forward velocity, from the total energy increment divided by the stroke period.

Depending on the forward velocity, different methods are required for estimating the power components in Eq. (1.30). If the forward velocity is high, the unsteady effects are small and quasi-steady assumptions can give good approximations. For slow forward velocities the vortex theory is more accurate, especially when one is estimating the induced power.

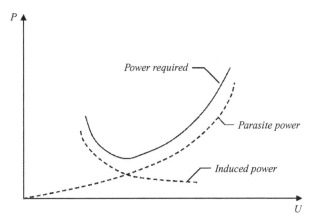

Figure 1.17. The two power components for a fixed-wing air vehicle, and the *power required*, as a result of adding these two components together. The *parasite power* curve represents the function $P = f(U^3)$ and the *induced power* curve $P = f(U^{-1})$.

Besides the components previously introduced, there is another component called the *inertial power* (P_{iner}), which is the power needed to move the wings and only the wings. The most important parameter when one is calculating this power is the moment of inertia I of the wing. Two main ways exist to obtain a low moment of inertia, namely, to keep the mass of the wing as low as possible and to concentrate the mass as much as possible near the axis of rotation. The *inertial power* is typically small under a medium to fast forward-flight condition and can be neglected (Norberg, 1976). However, for slow or hovering flight this power must be accounted for.

The total power (P_{tot}) required for flight is the sum of the total aerodynamic power and the inertial power:

$$P_{tot} = P_{aero} + P_{iner} = P_{ind} + P_{pro} + P_{par} + P_{iner}. \tag{1.31}$$

This is only the power required for flight and is not the same as the power input (Goldspink, 1977). Because the flight muscles are limited by their own mechanical efficiency, and all living animals are regulated by their own metabolism, the power input needed is higher than the total power required in Eq. (1.31).

The power required (P_{tot}) is strongly connected to the forward-flight speed. A common way of describing this relationship is by means of a power curve. For a fixed-wing air vehicle, the induced power is proportional to U^{-1} and the profile and parasite powers to U^3, the power required is given by

$$P \sim k_5 U^{-1} + k_6 U^3, \tag{1.32}$$

where k_5 and k_6 are constants.

When each power component is expressed as a function of velocity, $P = f(U^3)$ and $P = f(U^{-1})$, two curves can be plotted (see Figure 1.17). The solid curve in Figure 1.17 represents the power required for steady flight.

The most common power–flight-speed curve is the U-shaped curve, as suggested in Figure 1.17, which is further illustrated in Figure 1.18, in which there exists a

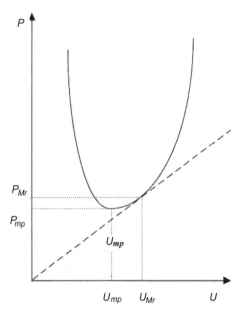

Figure 1.18. The U-shaped power curve for a fixed-wing aircraft. U_{mp} is the velocity for minimum power (P_{mp}) and U_{Mr} is the velocity for maximum range.

particular speed U_{mp}, where the power required has a minimum value. Also in Figure 1.18 is a straight dashed line. This line starts at the origin and tangents the U-curve at a certain point. The velocity at this point is the velocity for maximum range, U_{Mr}. When flyers migrate they need to cover a long distance for a given amount of energy, and therefore tend to fly at this velocity.

For a fixed wing, as shown by Lighthill (1977), U_{mp} and U_{Mr} are related by

$$U_{\mathrm{Mr}} = 1.32\, U_{\mathrm{mp}}. \tag{1.33}$$

For birds, the power curve is not necessarily U-shaped. Different researchers in the area of avian flight have come up with different shapes of the power curve (Alexander, 1997). The differences could be explained by the different ways approached by the researchers in, for example, the power components that have been considered and the muscle efficiencies that have been used. Nevertheless, as shown in Figure 1.19, the U-shaped power–flight-speed curve is indeed observed in natural flyers.

1.4   Concluding Remarks

In this chapter, we have offered an overview of the various low Reynolds number flyers, highlighting flight characteristics and scaling laws related to wingspan, wing area, wing loading, and vehicle size and weight.

The scaling laws indicate that, as a flyer's size reduces, it has to flap faster to stay in air, experiences lower wing loading, is capable of cruising slower, has a lower stall speed, and consequently can survive much better in a crash landing. In the meantime,

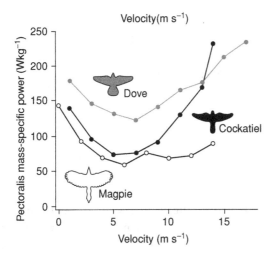

Figure 1.19. Pectoralis power as a function of flight velocity. Comparative mass-specific pectoralis power as a function of flight velocity in cockatiels, doves, and magpies. Bird silhouettes are shown to scale, digitized from a video (Tobalske et al., 2003).

as a flyer becomes smaller, its weight shrinks at a much faster rate, meaning that it can carry very little "fuel" and has to resupply frequently. Birds, bats, and insects apply different flapping patterns in hovering and forward flight to generate lift and thrust. Typically, in slow forward flight the reduced frequency and wing-beat amplitude tend to be high, resulting in highly unsteady flow structures. In fast forward flight the reduced frequency and the wing-beat amplitude tend to be low, and the wake often consists of a pair of continuous undulating vortex tubes or line vortices. Larger birds have relatively simple wingtip paths compared with those of smaller flyers. We have also discussed the power requirement associated with flight, including the *U*-shaped curve between specific power and flight speed.

As a flyer's size reduces, its operating Reynolds number becomes lower; accordingly, its wing in a *steady* stream produces a poorer lift-to-drag ratio. Coupling with a slower flight speed, a small flyer is substantially more influenced by the flight environment such as wind gust. To overcome these challenges, natural flyers flap their wings to enhance lift and improve maneuverability. The aerodynamics of fixed and flapping wings are discussed next.

CHAPTER TWO

Fixed, Rigid-Wing Aerodynamics

As already mentioned, there are several prominent features of MAV flight: (i) low Reynolds numbers (10^4–10^5), resulting in degraded aerodynamic performance, (ii) small physical dimensions, resulting in much reduced payload capabilities, as well as some favorable scaling characteristics including structural strength, reduced stall speed, and impact tolerance, (iii) low flight speed, resulting in an order one effect of the flight environment such as wind gust, and intrinsically unsteady flight characteristics. The preferred low Reynolds number airfoil shapes are different from those typically used for manned aircraft in thickness, camber, and AR. In this chapter, we discuss low Reynolds number aerodynamics and the implications of airfoil shapes, laminar–turbulent transition, and an unsteady free stream on the performance outcome.

Schmitz (1942) was among the first to investigate the aerodynamics for model airplanes in Germany, and he published his research in 1942. His work is often considered to be the first reported low-speed wind-tunnel research. However, before him, experimental investigations of low Reynolds number aerodynamics were conducted by Brown (1939) and by Weiss (1939), in the first two (and only) issues of *The Journal of International Aeromodeling*. Brown's experiments focused on curved-plate airfoils, made by using two circle arcs meet at maximum camber point of 8%, at varying locations. The wing test sections were all of 12.7 cm × 76.2 cm, giving an aspect ratio of 6. In all cases, the tests were conducted at a free-stream velocity of 94 cm/s. The Reynolds number, although not mentioned in Brown's study, is estimated to be about 8000.

It is hard to judge the quality of the measurements reported in these early works (see Figure 2.1). Nevertheless, their publications have clearly demonstrated the fact that a model airplane offered and continues to offer enthusiastic inquiries of many aspects related to the low Reynolds aerodynamics. Representative figures from Brown's experiments (Brown, 1939) are included here for us to gain a historical perspective.

Many papers have been published to improve our understanding, experimental database, and airfoil design guidance in the lower Reynolds number regime. For example, valuable insight has been offered by Liebeck (1992), Selig et al. (1995, 1996a, 1996b), and Hsiao et al. (1989). Liebeck (1992) addressed the laminar separation and airfoil design issues for Reynolds numbers Re between 2×10^5 and 2×10^6, and Hsiao

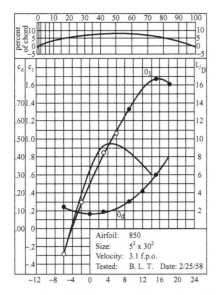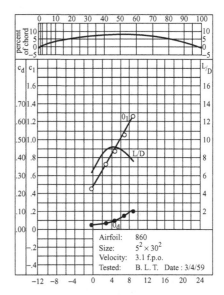

Figure 2.1. Low-speed aerodynamic tests reported by Brown for two airfoils (Brown, 1939). The chord was 12.7 cm and the free-stream velocity was 94 cm/sec.

et al. (1989) investigated the aerodynamic and flow structure of an airfoil, NACA 63_3-018, for *Re* between 3×10^5 and 7.74×10^5. Selig et al. covered a wide variety of airfoils to obtain basic aerodynamics data for *Re* between 6×10^4 and 3×10^5 (Selig et al., 1995, 1996b) and for *Re* between 4×10^4 and 3×10^5 (Selig et al., 1996a). In the following text, we discuss the various aerodynamics characteristics and fluid physics for *Re* between 100 and 10^6. Our main interest is on issues related to a Reynolds number of 10^5 or lower.

2.1 Laminar Separation and Transition to Turbulence

Figure 2.2 illustrates the aerodynamic performances and shapes of several representative airfoils under a steady-state free stream. A substantial reduction in the lift-to-drag ratio is observed as the Reynolds number becomes lower. The observed aerodynamic characteristics are associated with the laminar–turbulent transition process. For conventional manned aircraft wings, whose Reynolds numbers exceed 10^6, the flows surrounding them are typically turbulent, with the near-wall fluid capable of strengthening its momentum by means of energetic "mixing" with the free stream. Consequently flow separation is not encountered until the AoA becomes high. For low Reynolds number aerodynamics, the flow is initially laminar and is prone to separate even under a mild adverse pressure gradient. Under certain circumstances, as discussed next, the separated flow reattaches and forms a laminar separation bubble (LSB) while transitioning from a laminar to a turbulent state. Laminar separation can modify the effective shape of an airfoil and consequently influence the aerodynamic performance.

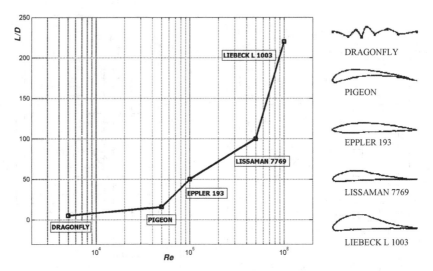

Figure 2.2. Aerodynamic characteristics of representative airfoils. Figure plotted based on data from Lissaman (1983).

The first documented experimental observation of a LSB was reported by Jones (1938). In general, under an adverse pressure gradient of sufficient magnitude, the laminar fluid flow tends to separate before becoming turbulent. After separation, the flow structure becomes increasingly irregular, and, beyond a certain threshold, it undergoes transition from laminar to turbulent. The turbulent mixing process brings high-momentum fluid from the free stream to the near-wall region, which can overcome the adverse pressure gradient, causing the flow to reattach.

The main features of a LSB are illustrated in Figure 2.3(a). After separation, the laminar flow forms a free-shear layer, which is contained between outer edge S″T″ of the viscous region and the mean dividing streamline ST′. Downstream of the transition point T, turbulence can entrain significant amount of high-momentum fluid through diffusion (Roberts, 1980), which enables the separated flow to reattach to the wall and form a turbulent free-shear layer. The turbulent free-shear layer is contained between lines T″R″ and T′R. The recirculation zone is bounded by the ST′R and STR.

Just downstream of the separation point, there is a "dead-fluid" region, where the recirculation velocity is significantly less than the free-stream velocity and the flow can be considered almost stationary. Because the free-shear layer is laminar and is less effective in mixing, the flow velocity between the separation and transition is virtually constant (Roberts, 1980). This is also reflected in the pressure distribution in Figure 2.3(b). The pressure "plateau" is a typical feature of the laminar part of the separated flow.

The dynamics of a LSB depends on the value of the Reynolds number, the pressure distribution, the geometry, the surface roughness, and the free-stream turbulence. An empirical rule given by Carmichael (1981) says that the Reynolds number, based on the free-stream velocity and the distance from the separation point to the

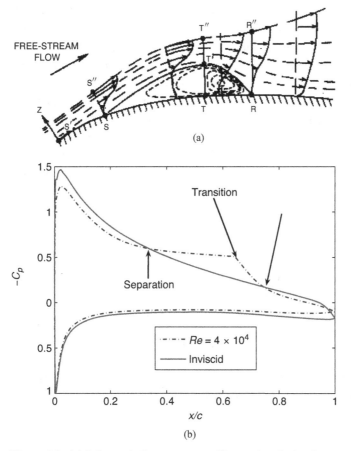

Figure 2.3. (a) Schematic flow structures illustrating the laminar–turbulent transition (Roberts, 1980) (copyright by AIAA). (b) Pressure distribution over an SD7003 airfoil, as predicted by XFOIL (Drela, 1989).

reattachment point, is approximately 5×10^4. It suggests that, if the Reynolds number is less than 5×10^4, an airfoil will experience separation without reattachment; on the other hand, a long separation bubble will occur if the Reynolds number is slightly higher than 5×10^4. This rule provides a general guide to predict the reattachment and should be used with caution. As we discuss later, the transition and the reattachment process is too complicated to be described by the Reynolds number alone.

As the Reynolds number decreases, the viscous damping effect increases, and it tends to suppress the transition process or delay reattachment. The flow will not reattach if

1. the Reynolds number is sufficiently low for the flow to completely remain laminar, or
2. the pressure gradient is too strong for the flow to reattach. Thus, without reattachment, a bubble does not form and the flow is then fully separated.

Based on its effect on pressure and velocity distribution, the LSB can be classified as either a short or long bubble (Tani, 1964). A short bubble covers a small portion of the airfoil and plays an insignificant rule in modifying the velocity and pressure distributions over an airfoil. In such a case, the pressure distribution closely follows its corresponding inviscid distribution except near the bubble location, where there is a slight deviation from the inviscid distribution. On the other hand, a long bubble covers a considerable portion of the airfoil and significantly modifies the inviscid pressure distribution and velocity peak. The presence of a long bubble leads to a decrease of lift and an increase of drag. Typically, a separation bubble has very steep gradients in the edge velocity u_e and momentum thickness θ at the reattachment point, resulting in jumps in Δu_e and $\Delta \theta$ over a short distance. For incompressible flow, the momentum thickness is defined as

$$\theta = \int_0^{\xi \to \infty} \frac{u}{U} \left(1 - \frac{u}{U} \right) d\xi, \tag{2.1}$$

where u is the streamwise velocity and U is the free-stream velocity. For flow over a flat plate, the momentum thickness is equal to drag force divided by ρU^2. If the skin friction is omitted, the correlation between these jumps can be expressed as

$$\frac{\Delta \theta}{\theta} \cong - (2 + H) \frac{\Delta u_e}{u_e}, \tag{2.2}$$

where H is the shape factor, defined as the ratio between the boundary-layer displacement thickness δ^* and the momentum thickness θ. The boundary-layer displacement is defined as

$$\delta^* = \int_0^{\xi \to \infty} \left(1 - \frac{u}{U} \right) d\xi. \tag{2.3}$$

Because of the change in flow structures, the shape factor H increases rapidly downstream of the separation point. Hence, according to correlation (2.2), the momentum-thickness jump is sensitive to the location of transition point in the separation bubble. Furthermore, because airfoil drag is directly affected by a momentum-thickness jump, an accurate laminar–turbulent transition model is important for drag prediction.

Figure 2.4 illustrates the behavior of an LSB in response to the Reynolds number. The analyses are based on the XFOIL code (Drela, 1989), which uses the thin-layer fluid flow model, assuming that the transverse length scale is much smaller than the streamwise length scale. At a fixed AoA, four flow regimes can be identified as the Reynolds number varies. As indicated in Figure 2.4, at $Re = 10^6$, on the upper surface there exists a short LSB, which affects the velocity distribution only locally. At an intermediate Reynolds number, e.g., $Re = 4 \times 10^4$, the short bubble bursts to form a long bubble. The peak velocity is substantially lower than that of the inviscid flow.

As the Reynolds number decreases to, e.g., $Re = 2 \times 10^4$, the velocity peak and circulation decrease further, reducing the pressure gradient after the suction peak. A weaker pressure gradient attenuates the amplification of disturbance in the laminar boundary layer, which delays the transition and elongates the free-shear

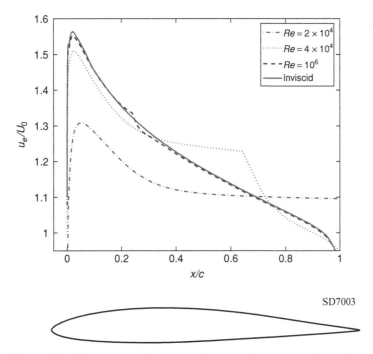

Figure 2.4. Streamwise velocity profiles over the upper surface of an SD7003 airfoil with varying Reynolds numbers, at a fixed AoA of $4°$. Inviscid as well as viscous flow solutions are shown. At $Re = 10^6$, a short bubble is observed; otherwise, the velocity distribution largely matches that of the inviscid model. At a lower Reynolds number, $Re = 4 \times 10^4$, a long bubble after bursting is observed, causing significant impact on the velocity distribution. Finally, at $Re = 2 \times 10^4$, a complete separation with no reattachment is noticed. Here u_e is the velocity at the boundary-layer edge parallel to the airfoil surface, and U_0 is the free-stream velocity. The results are based on computations made with XFOIL (Drela, 1989).

layer. At this Reynolds number, the separated flow no longer reattaches to the airfoil surface. The main structures are no longer sensitive to the exact value of the Reynolds number.

For a fixed Reynolds number, varying the AoA changes the pressure gradient aft of the suction peak and therefore changes the LSB. In this aspect, varying the AoA has the same effect on the LSB as changing the Reynolds number. Figure 2.5 illustrates that, at a fixed Reynolds number of 60,100, for the Eppler E374, a zigzag pattern appears in the lift–drag polar:

1. At a lower AoA, for example $2.75°$, there is a long bubble on the airfoil surface, which leads to a large drag.
2. When the AoA is increased (from $4.03°$ to $7.82°$), the adverse pressure gradient on the upper surface grows, which intensifies the Tollmien–Schlichting (TS) wave, resulting in an expedited laminar–turbulent transition process. A shorter LSB leads to more airfoil surface covered by the attached turbulent boundary-layer flow, resulting in a lower drag. This corresponds to the lift–drag polar's left turn.

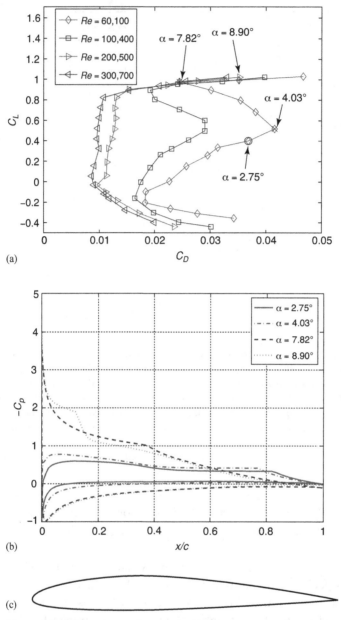

(a)

(b)

(c)

Figure 2.5. Lift-to-drag polars of the Eppler E374 at different chord Reynolds numbers: (a) lift-to-drag polars at different Reynolds numbers; (b) pressure coefficient distributions at different AoAs for $Re = 60{,}100$. Data are computed with XFOIL (Drela, 1989); (c) E374 airfoil.

3. When the AoA is further increased (beyond $7.82°$), the separated flow quickly experiences transition; however, with a massive separation, the turbulent diffusion can no longer make the flow reattach, and the drag increases substantially with little changes in lift.

The previously described zigzag pattern of the lift–drag polar is a noticeable feature of low Reynolds number aerodynamics. As illustrated in Figure 2.5, at a sufficiently high Reynolds number, the polar exhibits the familiar C-shape.

Earlier experimental investigations on low Reynolds number aerodynamics were reviewed by Young and Horton (1966). Carmichael (1981) further reviewed theoretical and experimental results of various airfoils with Reynolds numbers spanning from 10^2 to 10^9. In particular, the near-surface flow and aerodynamic loads of a wing at Reynolds numbers in the range 10^4–10^6 were studied by many investigators. Crabtree (1957) studied the formation of short and long LSBs on thin airfoils. Consistent with the preceding discussion of the two types of separation bubbles, he suggested that the long bubble directly influences aerodynamic characteristics whereas the short one serves as an agent for initiating a turbulent boundary layer. Numerous further investigations have been reported in the past two decades on the interplay between near-wall flow structures and the aerodynamic performance. For example, Huang et al. (1996) studied the aerodynamic performance versus the surface-flow mode at different Reynolds numbers. Hillier and Cherry (1981) and Kiya and Sasaki (1983) studied the influence of the free-stream turbulence on the separation bubble along the side of a blunt plate with right-angled corners and found that the bubble length, sizes of vortices in the separating region, and the level of the suction peak pressure can all be well correlated with the turbulence outside the shear layer and near the separation point.

2.1.1 *Navier–Stokes Equation and the Transition Model*

To perform practical laminar- and turbulent-flow computations in the Reynolds number range typically used by the low Reynolds number flyers, the constant-property Navier–Stokes equations adequately model the fluid physics:

$$\frac{\partial u_i}{\partial x_i} = 0, \tag{2.4}$$

$$\frac{\partial u_i}{\partial t} + \frac{\partial}{\partial x_j}(u_i u_j) = -\frac{1}{\rho}\frac{\partial p}{\partial x_i} + \nu\frac{\partial^2}{\partial x_j^2}(u_i), \tag{2.5}$$

where u_i are the mean flow velocities and ν is the kinematic viscosity. For turbulent flows, turbulent closures are needed if one is solving the ensemble-averaged form of the Navier–Stokes equations. Numerous closure models have been proposed in the literature (Wilcox, 2000). Here we present the two-equation k–ω turbulence model (Wilcox, 2000) as an example. For clarity, the turbulence model is written in Cartesian coordinates as follows:

$$\frac{\partial k}{\partial t} + \frac{\partial(u_j k)}{\partial x_j} = \tau_{ij}\frac{\partial u_i}{\partial x_j} - \beta^*\omega k + \frac{\partial}{\partial x_j}\left[(\nu + \sigma^*\nu_T)\frac{\partial k}{\partial x_j}\right], \tag{2.6}$$

$$\frac{\partial \omega}{\partial t} + \frac{\partial(u_j\omega)}{\partial x_j} = \frac{\alpha\omega}{k}\tau_{ij}\frac{\partial u_i}{\partial x_j} - \beta\omega^2 + \frac{\partial}{\partial x_j}\left[(\nu + \sigma\nu_T)\frac{\partial \omega}{\partial x_j}\right], \tag{2.7}$$

where

$$v_T = \alpha^* k / \omega, \quad \tau_{ij} = 2v_T S_{ij} - 2/3 k \delta_{ij}, \quad S_{ij} = \frac{1}{2}\left(\frac{\partial u_i}{\partial x_j} + \frac{\partial u_j}{\partial x_i}\right), \quad (2.8)$$

$$\alpha^* = \frac{\alpha_0^* + Re_T/R_k}{1 + Re_T/R_k}, \quad \alpha = \frac{13}{25}\frac{\alpha_0 + Re_T/R_\omega}{1 + Re_T/R_\omega}\frac{1}{\alpha^*}, \quad \beta^* = \frac{9}{100}\frac{4/15 + (Re_T/R_\beta)^4}{1 + (Re_T/R_\beta)^4},$$
$$(2.9)$$

$$Re_T = \frac{k}{\omega v}, \quad \beta = \frac{9}{125}, \quad \sigma^* = \sigma = \frac{1}{2}, \quad \alpha_0^* = \frac{1}{3}\beta, \quad \alpha_0 = \frac{1}{9}, \quad (2.10)$$

$$R_\beta = 8, \quad R_k = 6, \quad R_\omega = 2.95. \quad (2.11)$$

For the preceding equations, k is the turbulent kinetic energy, ω is the dissipation rate, v_T is the turbulent kinematic eddy viscosity, Re_T is the turbulent Reynolds number, and $\alpha_0, \beta, R_\beta, R_k$, and R_ω are model constants. To solve for the transition from laminar to turbulent flow, the incompressible Navier–Stokes equations are coupled with a transition model.

The onset of laminar–turbulent transition is sensitive to a wide variety of disturbances, such as pressure gradient, wall roughness, free-stream turbulence, acoustic noise, and thermal environment. A comprehensive transition model considering all these factors currently does not exist. Even if we limit our focus on free-stream turbulence, it is still a challenge to give an accurate mathematical description. Overall, approaches of transition prediction can be categorized as (i) empirical methods and methods based on linear stability analysis, such as the e^N method (Drela, 1989), (ii) linear or nonlinear parabolized stability equations (Herbert, 1997), and (iii) large-eddy simulation (LES) (Lesieur and Metais, 1996) or direct numerical simulation (DNS) methods (Moin and Mahesh, 1998).

Empirical methods have also been proposed to predict transition in a separation bubble. For example, Roberts (1980), Davis et al. (1987), and Volino and Bohl (2004) developed models based on local turbulence levels; Mayle (1991), Praisner and Clark (2004), and Roberts and Yaras (2005) tested concepts by using the local Reynolds number based on the momentum thickness. These models use only one or two local parameters to predict the transition points and hence often oversimplify the downstream factors such as pressure gradient, surface geometry, and surface roughness. For attached flow, Wazzan et al. (1979) proposed a model based on the shape factor H. Their model gives a unified correlation between the transition point and Reynolds number for a variety of problems. For separated flow, however, no similar models exist, in part because of the difficulty in estimating the shape factor.

Among the approaches using linear stability analysis, the e^N method is widely adopted (Smith and Gamberoni, 1956; Van Ingen, 1956). It solves the Orr–Sommerfeld equation to evaluate the local growth rate of unstable waves based on velocity and temperature profiles over a solid surface. Its successful application is exemplified in the popularity of airfoil analysis software such as XFOIL (Drela, 1989). XFOIL uses the steady Euler equations to represent the inviscid flows, a two-equation integral formulation based on dissipation closure to represent boundary layers and wakes and the e^N method to tackle transition. The concept of coupling a

Reynolds-averaged Navier–Stokes (RANS) solver with e^N method to predict transition has been practiced by Radespiel et al. (1991), Stock and Haase (1999), and He et al. (2000). A recent application of this approach for low Reynolds number applications can be found in the work of Yuan et al. (2005) and Lian and Shyy (2006).

The e^N method is based on the following assumptions: (i) the velocity and temperature profiles are essentially two dimensional (2D) and steady; (ii) the initial disturbance is infinitesimal; and (iii) the boundary layer is thin. Even though in practice the e^N method has been extended to study the three-dimensional (3D) flow, strictly speaking, such flows do not meet the preceding conditions. Furthermore, even in 2D flow, not all these assumptions can be satisfied (Zheng et al., 1998). Nevertheless, the e^N method remains a practically useful approach for engineering applications.

The advancement in turbulence modeling has offered alternative approaches for transition prediction. For example, Wilcox devised a low Reynolds number k–ω turbulence model to predict transition (Wilcox, 1994). One of his objectives is to match the minimum critical Reynolds number beyond which the TS wave begins forming in the Blasius boundary-layer context. However, this model fails if the separation-induced transition occurs before the minimum Reynolds number, as frequently occurs in a separation-induced transition. Holloway et al. used unsteady RANS equations to study the flow separation over a blunt body for the Reynolds number range 10^4–10^7 (Holloway et al., 2004). It is observed that the predicted transition point can be too early even for a flat plate flow case, as illustrated by Dick and Steelant (1997). In addition, Dick and Steelant (1996) and Suzen and Huang (2000) incorporated the concept of an intermittency factor to model the transitional flows. One can achieve this either by using conditional-averaged Navier–Stokes equations or by multiplying the eddy viscosity by the intermittency factor. In either approach, the intermittency factor is solved based on a transport equation, aided by empirical correlations. Mary and Sagaut (2002) studied the near-stall phenomena around an airfoil by using LES, and Yuan et al. (2005) studied transition over a low Reynolds number airfoil by using LES. The major challenge for the computations of transitional flow by use of LES is the artificial triggering of transition by a pointwise input of turbulent kinetic energy.

2.1.2 The e^N Method

In the following discussion, we offer a more detailed presentation of the e^N method because it forms the basis for low Reynolds number aerodynamics predictions and has proven to be useful for engineering applications. As already mentioned, the e^N method is based on linear stability analysis, which states that transition occurs when the most unstable TS wave in the boundary layer has been amplified by certain factors. Given a velocity profile, one can determine the local disturbance growth rate by solving the Orr–Sommerfeld eigenvalue equations. Then, one calculates the amplification factor by integrating the growth rate, usually the spatial growth rate, starting from the point of neutral stability. The *Transition Analysis Program System* (TAPS) by Wazzan and coworkers (Wazzan et al., 1968) and the *COSAL* program by Malik

(1982) can be used to compute the growth rate for a given velocity profile. Schrauf also has developed a program called *Coast3* (Schrauf, 1998). However, it is very time consuming to solve the eigenvalue equations. An alternative approach has been proposed by Gleyzes et al. (1985) who found that the integrated amplification factor $\tilde{n}$ can be approximated by an empirical formula as follows:

$$\tilde{n} = \frac{d\tilde{n}}{dRe_\theta} (H) \left[Re_\theta - Re_{\theta_0}(H) \right], \tag{2.12}$$

where, as previously defined, θ is the boundary-layer momentum thickness, Re_θ is the momentum-thickness Reynolds number, Re_{θ_0} is the critical Reynolds number that we will define later, and H is the shape factor previously discussed. With this approach, one can approximate the amplification factor with a reasonably good accuracy without solving the eigenvalue equations. For similar flows, the amplification factor $\tilde{n}$ is determined by the following empirical formula:

$$\frac{d\tilde{n}}{dRe_\theta} = 0.01 \left\{ [2.4H - 3.7 + 2.5 \tanh(1.5H - 4.65)]^2 + 0.25 \right\}^{1/2}. \tag{2.13}$$

For nonsimilar flows, i.e., those that cannot be treated by similarity variables by use of the Falkner–Skan profile family (White, 1991), the amplification factor with respect to the spatial coordinate ξ is expressed as

$$\frac{d\tilde{n}}{d\xi} = \frac{d\tilde{n}}{dRe_\theta} \frac{1}{2} \left(\frac{\xi}{u_e} \frac{du_e}{d\xi} + 1 \right) \frac{\rho u_e \theta^2}{u_e \xi} \frac{1}{\theta}. \tag{2.14}$$

An explicit expression for the integrated amplification factor then becomes

$$\tilde{n}(\xi) = \int_{\xi_0}^{\xi} \frac{d\tilde{n}}{d\xi} d\xi, \tag{2.15}$$

where ξ_0 is the point where $Re_\theta = Re_{\theta_0}$, and the critical Reynolds number is expressed by the following empirical formula:

$$\log_{10} Re_{\theta_0} = \left(\frac{1.415}{H-1} - 0.489 \right) \tanh \left(\frac{20}{H-1} - 12.9 \right) + \frac{3.295}{H-1} + 0.44. \tag{2.16}$$

Once the integrated growth rate reaches the threshold N, flow becomes turbulent. To incorporate the free-stream turbulence level effect, Mack (1977) proposed the following correlation between the free-stream intensity T_i and the threshold N:

$$N = -8.43 - 2.4 \ln(T_i), \quad 0.0007 \le T_i \le 0.0298. \tag{2.17}$$

Care should be taken in using such a correlation. The free-stream turbulence level itself is not sufficient to describe the disturbance because other information, such as the distribution across the frequency spectrum, should also be considered. The so-called "receptivity," i.e., how the initial disturbances within the boundary layer are related to the outside disturbances, is a critically important issue. Actually, we can determine the N factor only if we know the "effective T_i," which can be defined only through a comparison of the measured transition position with calculated amplification ratios (Van Ingen, 1995).

A typical procedure to predict transition point with coupled RANS equations and the e^N method is as follows: The Navier–Stokes equations together with a turbulence model are first solved without invoking the turbulent production terms, for which the flow is essentially laminar; the amplification factor $\tilde{n}$ is integrated based on Eq. (2.12) along the streamwise direction; once the value reaches the threshold N, the production terms are activated for the post-transition computations. After the transition point, flow does not immediately become fully turbulent; instead, the process toward full turbulence is a gradual process. This process can be described with an intermittency function, allowing the flow to be represented by a combination of laminar and turbulent structures. With the intermittency function, an *effective eddy viscosity* is used in the turbulence model, and it can be expressed as follows:

$$\nu_{Te} = \gamma \nu_T, \qquad (2.18)$$

where γ is the intermittency function, ν_{Te} is the effective eddy viscosity, and ν_T is the eddy viscosity.

In the literature a variety of intermittency distribution functions is proposed. For example, Cebeci (1988) has presented such a function by improving on a model previously proposed by Chen and Thyson (1971) for the Reynolds number range of 2.4×10^5 to 2×10^6 with an LSB. However, no model is available when the Reynolds number is lower than 10^5. Lian and Shyy (2006) suggested that, for separation-induced transition at such a low Reynolds number regime, the intermittency distribution is largely determined by the distance from the separation point to the transition point, and the shorter the distance, the quicker flow becomes turbulent. Also, previous work suggests that the flow property at the transition point will also be important. From the available experimental data and our simulation, they proposed the following model (Lian and Shyy, 2006):

$$\gamma = \begin{cases} 1 - \exp\left(-\left\{\exp\left[\frac{\max(H_T - 2.21, 0)}{20}\right]^2 - 1\right\}\left(\frac{x - x_T}{x_T - x_S}\right) Re_{\theta T}\right) & (x \geq x_T) \\ 0 & (x < x_T) \end{cases}, \qquad (2.19)$$

where x_T is the transition onset position, x_S is the separation position, H_T is the shape factor at the transition point, and $Re_{\theta T}$ is the Reynolds number based on the momentum thickness at the transition point.

2.1.3 *Case Study: SD7003*

Lian and Shyy (2006) studied the Reynolds number effect with a Navier–Stokes equation solver augmented with the e^N method. The lift and drag coefficients of the SD7003 airfoil versus AoA are plotted in Figure 2.6. The good agreement between the numerical results (Lian and Shyy, 2006) and the experimental measurements by Ol et al. (2005) and Selig et al. (1995) can be seen vividly. Both the simulation and the measurement by Ol et al. (2005) predict that the maximum lift coefficient happens at $11°$. Close to the stall AoA, the simulations overpredict the lift coefficients.

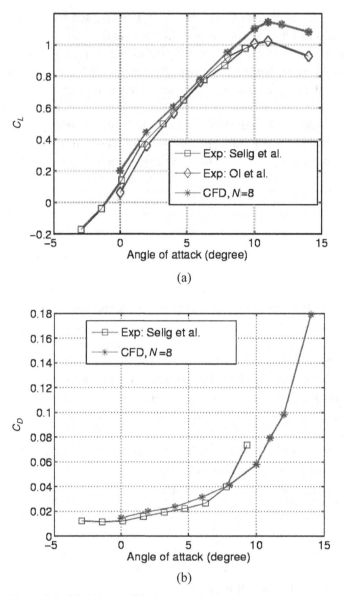

Figure 2.6. (a) Lift and (b) drag coefficients vs. the AoA for the SD7003 airfoil at $Re = 6 \times 10^4$ (Lian & Shyy, 2006). CFD, computational fluid dynamics.

As the AoA increases, as illustrated in Figure 2.7, the adverse pressure gradient downstream of the point of the suction peak becomes stronger and the separation point moves toward the leading edge. The strong pressure gradient amplifies the disturbance in the separation zone and prompts transition. As the turbulence develops, the increased entrainment causes reattachment. At an AoA of 2°, the separation position is at around 37% of the chord length and transition occurs at 75% of the chord length. A long LSB forms. The plateau of the pressure distribution shown in Figure 2.7(a) is characteristic of such a long LSB. It is also

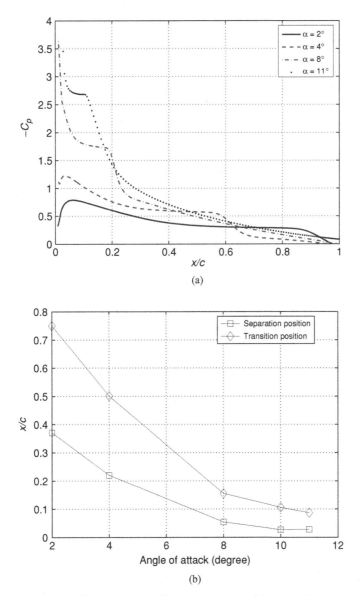

Figure 2.7. (a) Pressure coefficients vs. AoA, (b) separation and transition position versus the AoA for the SD7003 airfoil at $Re = 6 \times 10^4$ (Lian and Shyy, 2006).

noticed from Figure 2.7(b) that the bubble length decreases with an increase in the AoA.

Lian and Shyy (2006) compared the computed shear stress with the experimental measurement by Radespiel et al. (2006), utilizing a low-turbulence wind tunnel and a water tunnel because of the low-turbulence nature. Radespiel et al. (2006) suggest that large values of the critical N factor should be appropriate. As shown in Figure 2.8, the simulation by Lian and Shyy (2006) with $N = 8$ shows good agreement with measurement in terms of transition position, reattachment position, and vortex

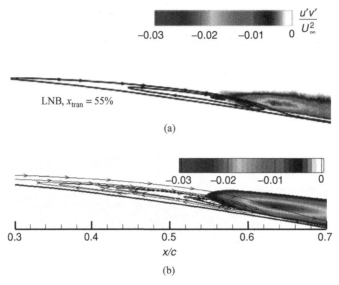

Figure 2.8. Streamlines and turbulent shear stress for $\alpha = 4°$: (a) experimental measurement by Radespiel et al. (2006), (b) numerical simulation with $N = 8$ by Lian and Shyy (2006).

core position. It should be noted that, in the experiment, the transition location is defined as the point where the normalized Reynolds shear stress reaches 0.1% and demonstrates a clearly visible rise. The transition point in the simulation is defined as the point where the most unstable TS wave has amplified over a factor of e^N. This definition discrepancy may cause some problems when we compare the transition position. In any event, overall, simulations typically predict noticeably lower shear-stress magnitude than the experimental measurement.

As the AoA increases, both the separation and the transition positions move upstream, and the bubble shrinks. The measurements at $\alpha = 8°$ and $11°$ are performed in the water tunnel with a measured free-stream turbulence intensity of 0.8%. At $\alpha = 8°$ the simulation by Lian and Shyy (2006) predicts that the flow goes though transition at 15% of the chord, which is close to the experiment measurement of 14%. The bubble covers approximately 8% of the airfoil upper surface. The computational and experimental results for the AoA of $\alpha = 8°$ are shown in Figure 2.9. With an AoA of $\alpha = 11°$, the airfoil is close to stall. The separated flow requires a greater pressure recovery in the laminar bubble for reattachment. Lian and Shyy (2006) predict that flow separates at 5% of the chord, and the separated flow quickly reattaches after it experiences transition at 7.5% chord position whereas the experiment showed transition occurred at 8.3%. This quick reattachment generally represents the transition-forcing mechanism. Comparison shows that the computed Reynolds shear stress matches the experiment measurement well (Figure 2.10).

For low Reynolds number airfoils, the chord Reynolds number is a key parameter used to characterize the overall aerodynamics. Between the separation position and the transition position, as shown in Figure 2.11(a), the shape factor H and the

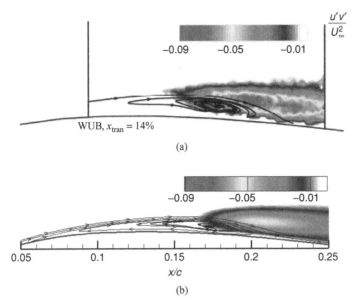

$\dfrac{u'v'}{U_\infty^2}$

−0.09   −0.05   −0.01

WUB, $x_{\text{tran}} = 14\%$

(a)

−0.09   −0.05   −0.01

0.05   0.1   0.15   0.2   0.25

x/c

(b)

Figure 2.9. Streamlines and turbulent shear stress for $\alpha = 8°$: (a) experimental measurement by Radespiel et al. (2006), (b) numerical simulation with $N = 3$ by Lian and Shyy (2006).

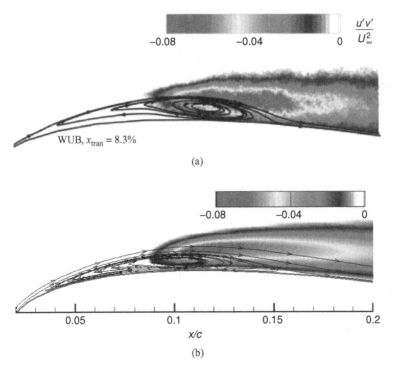

$\dfrac{u'v'}{U_\infty^2}$

−0.08   −0.04   0

WUB, $x_{\text{tran}} = 8.3\%$

(a)

−0.08   −0.04   0

0.05   0.1   0.15   0.2

x/c

(b)

Figure 2.10. Streamlines and turbulent shear stress for $\alpha = 11°$: (a) experimental measurement by Radespiel et al. (2006), (b) numerical simulation with $N = 3$ by Lian and Shyy (2006).

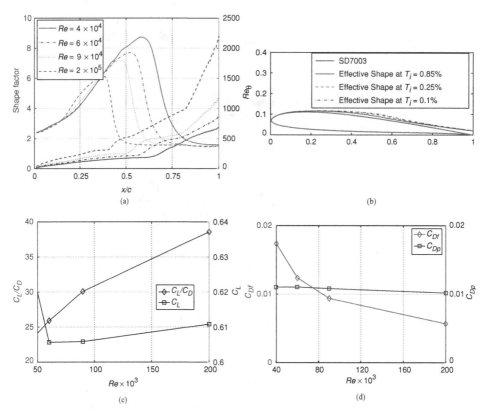

Figure 2.11. Reynolds number effect on the LSB profile and aerodynamic performance $\alpha = 4°$ for the SD7003 airfoil (Lian and Shyy, 2006): (a) shape factor and momentum-thickness-based Reynolds number, (b) effective airfoil shape, (c) lift-to-drag ratio, (d) drag coefficient.

momentum-thickness-based Reynolds number increase with the chord Reynolds number. As shown in Figure 2.11(b) the effective airfoil shape, which is the superimposition of the airfoil and the boundary-layer displacement thickness, at $Re = 4 \times 10^4$ has the largest camber. This helps explain why the largest lift coefficient is obtained at $Re = 4 \times 10^4$ (Figure 2.11(c)). The camber decreases significantly when the Reynolds number increases from 4×10^4 to 6×10^4 but does not show considerable change when the Reynolds number increases further. Therefore one does not observe much increase in the lift coefficient even though the LSB length is shorter at higher Reynolds numbers. One can conclude from Figure 2.11(d) that the enhancement of lift-to-drag ratio is mainly due to the reduction of friction drag at high Reynolds numbers. As the Reynolds number increases, the form drag does not vary as much as the friction drag.

2.2   Factors Influencing Low Reynolds Number Aerodynamics

With the influence of flow separation and laminar–turbulent transition, the preferred airfoil shapes in the low Reynolds number regime are different from those in the high Reynolds number regime. Furthermore, in addition to the Reynolds number, the airfoil camber, thickness, surface smoothness, freestream unsteadiness, and the AR all play important roles in determining the aerodynamic performance of a low Reynolds number flyer. These factors will be discussed in the following.

2.2.1 $Re = 10^3 - 10^4$

Okamoto et al. (1996) experimentally studied the effects of wing camber on wing performance with Reynolds numbers as low as 10^3–10^4. In their experiment, rectangular wings with an AR of 6, constructed from aluminum foil or balsa wood, are used. Figure 2.12 illustrates the effects of camber on the aerodynamic characteristics. As the camber increases, the lift coefficient slope and the maximum lift coefficient increase as well. The increase in camber pushes both the maximum lift coefficient and the maximum lift-to-drag ratio to a higher AoA. More interesting, the 3% camber airfoil shows stall-resisting tendency, with the lift just leveling off above an AoA of 10°. Apart from the disadvantage of a high drag coefficient, the low-camber airfoil does have the advantage in that it is less sensitive to the AoA and therefore does not require sophisticated steering.

Sunada et al. (2002) compared wing characteristics at a Reynolds number of 4×10^3. They fabricated rectangular wings with an AR of 7.25. Representative wings are shown in Figure 2.13. Among the 20 wings tested, they concluded that the wing performance can be improved with a modest camber of around 5%. Figure 2.14 shows the lift and drag coefficients versus the AoA. In their work, at $Re = 4 \times 10^3$, the effect of camber on aerodynamics is similar to that reported by Okamoto et al. (1996). In either experiment, the lift curve slope increases with the camber; a higher-camber wing has a higher-stall AoA, and generally a larger drag coefficient than a lower-camber wing at the same AoA. If we further compare wings of comparable cambers, we notice that they have almost the same stall angle. Sunada et al. (2002) further investigated the impact of the maximum-camber location, shown in Figure 2.15. They found that both lift and drag coefficients increase as the position of the maximum camber approaches the trailing edge. In terms of lift-to-drag ratio, the maximum value is obtained when the maximum camber is positioned at 25% chord.

Okamoto et al. (1996) also studied the effects of airfoil thickness. They found that the wing aerodynamic characteristics deteriorate as the wing thickness increases (Figure 2.16).

In contrast to conventional airfoils, which are smooth and streamlined, insect airfoils exhibit rough surfaces such as the cross-sectional corrugations of dragonfly wings (shown in Figure 2.17) or scales on the wing surface (butterfly and moth). Evidence has shown that the corrugated wing configuration bears both structural and aerodynamic benefits to the dragonflies. First, it is of critical importance to the stability of its ultralight construction. Second, in visualizing experiments using corrugated wings, Newman et al. (1977) and Buckholz (1986) showed that this geometry helps improve aerodynamic performance. The reason, as suggested by Kesel, is that vortices fill the profile valleys formed by these bends and therefore smooth the profile geometry

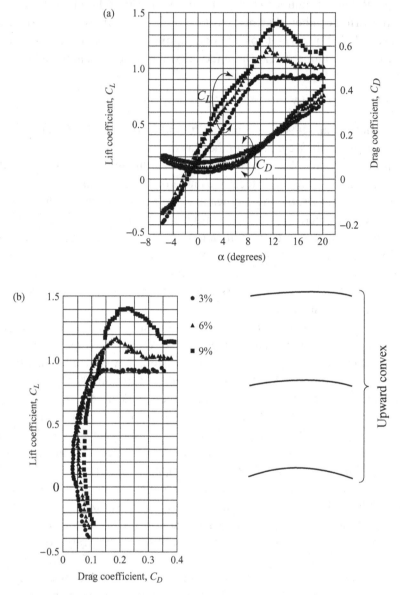

Figure 2.12. Effects of circular camber on the aerodynamic characteristics of a rectangular model wing made from aluminum foil, with a thickness of 0.3 mm and a chord length of 30 mm. Each symbol refers to a different camber, as shown in the panel on the right-hand side of the figure. (a) C_L and C_D vs. AoA, (b) polar curve. Adopted from Okamoto et al. (1996).

(Kesel, 1998). Kesel (2000) compared the aerodynamic characteristics of dragonfly wing sections with conventionally designed airfoils and flat plates at Reynolds numbers of 7.88×10^3 and 10^4. She concluded that corrugated airfoils, such as those seen in dragonflies (Figure 2.17), have very low drag coefficients closely resembling those of flat plates, whereas the lift coefficients are much higher than those of flat plates.

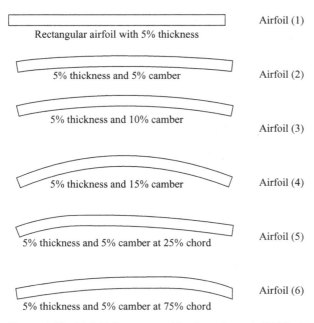

Figure 2.13. Airfoil shapes tested by Sunada et al. (2002). (Redrawn from the original reference with permission.)

She also investigated the performance of the airfoil by simply filling the valleys with solid materials (as illustrated in Figure 2.18). Figure 2.19 highlights the key features of the lift and drag values versus AoAs, between the three natural and filled airfoil profiles. Figure 2.20 shows the corresponding lift–drag polar. These plots, taken from Kesel (2000), show less favorable aerodynamic performances of the filled airfoils. Therefore it is clear that the performance of such a corrugated airfoil is influenced by its "effective" shape, characterized by the viscous effects, as previously discussed. In particular, the viscosity and associated vortical structures result in an airfoil with cambered geometry (Kesel, 2000).

2.2.2 $Re = 10^4$–10^6

Shyy et al. (1999b) evaluated the aerodynamics between the chord Reynolds numbers of 7.5×10^4 and 2×10^6 by using the XFOIL code (Drela, 1989) for two conventional airfoils, NACA 0012 and CLARK-Y, and two low Reynolds number airfoils, S1223 (Selig and Maughmer, 1992) and an airfoil modified from S1223, which is called UF (Figure 2.21). Figures 2.22 and 2.23 show the power index, $C_L^{3/2}/C_D$, and lift-to-drag ratio, C_L/C_D, plots at three Reynolds numbers, 7.5×10^4, 3×10^5, and 2×10^6. It is noted that, for steady-state flight, the power required for maintaining a fixed-wing vehicle in air is

$$P = W \left(C_D / C_L^{3/2} \right) \sqrt{\frac{2}{\rho} \frac{W}{S}}, \tag{2.20}$$

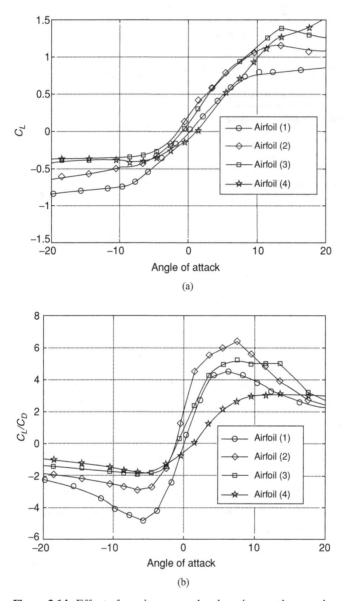

Figure 2.14. Effect of maximum-camber location on the aerodynamic characteristics at $Re = 4 \times 10^3$. Redrawn from Sunada et al. (2002) with permission.

where P and W are the required power and vehicle weight, respectively. For all airfoils, the C_L/C_D ratio exhibits a clear Reynolds number dependency. For Re varying between 7.5×10^4 and 2.0×10^6, C_L/C_D changes by a factor of 2–3 for the airfoils tested. Except for the UF airfoil, which is very thin, the range of AoA within which aerodynamics is satisfactory becomes narrower as the Reynolds number decreases. Clearly, the camber is important. NACA 0012, with 0% camber, and CLARK-Y, with 3.5% camber, yield a less satisfactory performance under all three Reynolds numbers. S1223 and UF, both with 8.89% camber, perform better. Finally, NACA 0012,

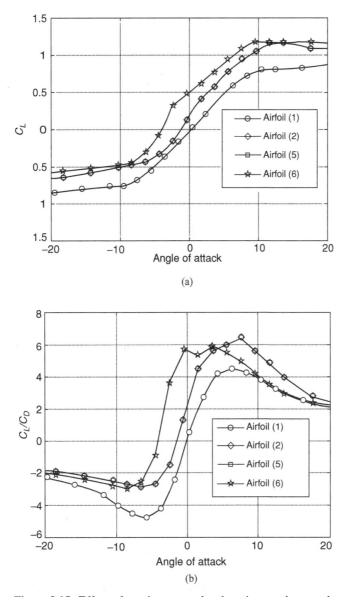

Figure 2.15. Effect of maximum-camber location on the aerodynamic characteristics at $Re = 4 \times 10^3$. Redrawn from Sunada et al. (2002) with permission.

CLARK-Y, and S1223 all have a maximum thickness of about $0.12c$. The UF airfoil, on the other hand, is considerably thinner, with a maximum thickness of $0.06c$. It is interesting to compare the Reynolds number effect. At $Re = 2.0 \times 10^6$, S1223 and UF have comparable peak performances in terms of $C_L^{3/2}/C_D$ and C_L/C_D; however, S1223 exhibits a wider range of acceptable AoAs. At $Re = 7.5 \times 10^4$, the situation is quite different. UF, the thinner airfoil with identical camber, exhibits a substantially better aerodynamic performance while maintaining a comparable range of acceptable AoAs. This is consistent with the findings of Okamoto et al. (1996), discussed previously.

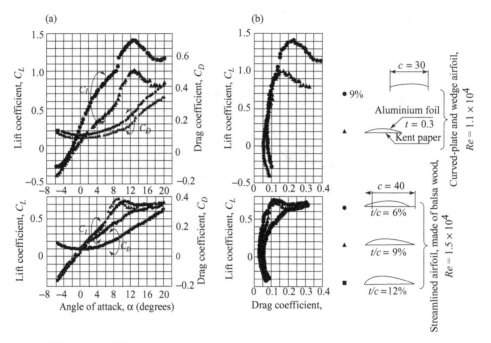

Figure 2.16. Effects of thickness on the aerodynamic characteristics of a curved-section model wing (camber 9%). Each symbol refers to a different airfoil shape as shown in the panel on the right-hand side; c, chord length; t, thickness; Re, Reynolds number. All dimensions are given in millimeters. (a) C_L and C_D vs. the AoA, (b) polar curve. Adopted from Okamoto et al. (1996).

2.2.3 *Effect of Free-Stream Turbulence*

When both the AoA and chord Reynolds number are fixed, increasing the free-stream turbulence level prompts earlier transition. The aerodynamic characteristics under different turbulence intensities were investigated by Lian and Shyy (2006). The lift and drag coefficients from their research are shown in Figure 2.24. At $\alpha = 4°$, there is no noticeable difference in the lift and drag coefficients among the five tested turbulence levels. This seemingly contradicts the pressure coefficient plot in Figure 2.25 because the integrated area between $C_p = 0$ and C_p distribution at $T_i = 0.85\%$ is smaller than that between $C_p = 0$ and the pressure coefficient distribution at $T_i = 0.07\%$. However, the integrated area is not linearly proportional to the lift because of the airfoil curvatures.

At $\alpha = 8°$, there is a drastic decrease in the lift coefficient and an increase in the drag coefficient when T_i decreases to 0.07%. Analysis of the flow structure reveals that, at such a low-turbulence level, the flow fails to reattach after its initial separation. This separation bubble causes the lift coefficient to drop by 10% and the drag coefficient to increase by more than 150%. Similar conclusions can also be drawn for the case of $\alpha = 11°$.

In general, with the increase of the free-stream turbulence level, the LSB becomes thinner and shorter. This is clearly shown in Figure 2.26. From the same figure it can

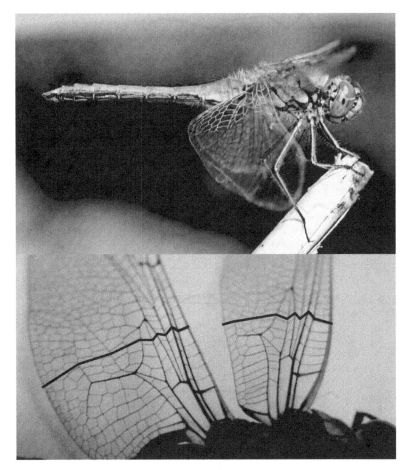

Figure 2.17. Dragonfly wings exhibit both flexibility and anisotropic, corrugated structures. In the lower picture, shown on the left-hand side is the hindwing and on the right-hand side is the forewing.
This figure is available in color for download from www.cambridge.org/9780521204019

also be seen that the shear stress decreases with the turbulence level. Because of the viscous effect, the boundary layer and the LSB change the effective shape of the airfoil. As shown in Figure 2.27, the free stream with a higher turbulence level results in a relatively thinner effective airfoil than that with a lower turbulence level.

O'Meara and Mueller (1987) experimentally studied the effects of free-stream turbulence on the separation bubble characteristics of the NACA 66_3–018 airfoil. They reported that, as the disturbance level is increased, the bubble is reduced in both length and thickness, which is consistent with the observations from Figures 2.26 and 2.27. As we will discuss later, the effects of increasing the disturbance level resemble the effects of increasing the chord Reynolds number. O'Meara and Mueller (1987) also reported that the suction peak grows in absolute magnitude with the disturbance level. However, as shown in Figure 2.25, the pressure peak over the SD7003 airfoil is not sensitive to the disturbance level. These two conclusions are drawn based on different test cases, in which the bubble size and Reynolds number are quite

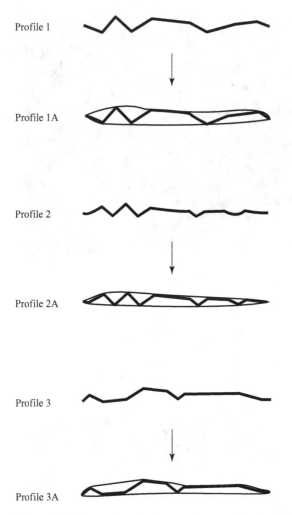

Figure 2.18. Geometry of wing profiles used in the study of Kesel (2000). Profiles 1, 2, and 3 are constructed with measurements taken from a dragonfly wing. Profiles 1A, 2A, and 3A are built by connection of the peaks of the respective cross sections.

different. The results in Figure 2.28 are obtained at a chord Reynolds number of 1.4×10^5; occupying around 7% of chord length, the bubble is short, and, as previously discussed, only locally affects the pressure distribution. On the other hand, in the test of Lian and Shyy (2006), the bubble covers more than 30% of the upper surface at the Reynolds number of 6×10^4 and $\alpha = 4°$, and the bubble falls into the long-bubble category. This hypothesis is further confirmed by the fact that, at a Reynolds number of 6×10^4 and $\alpha = 8°$, wherein the bubble is 8% of the chord, the pressure peak magnitude does increase with an increase in the disturbance level.

The effects of free-stream turbulence on lift and drag performances of a Lissaman 7769 airfoil are presented by Mueller et al. (Mueller et al., 1983). As shown in Figure 2.29, the hysteresis characteristics of the lift and the drag coefficients can be

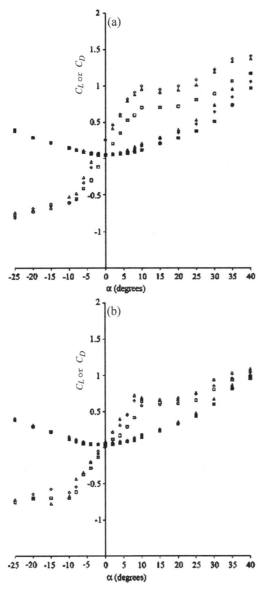

Figure 2.19. Comparison of the lift (open symbols) and drag (filled symbols) coefficients versus the AoAs, at $Re = 10,000$, for (a) three natural and (b) filled profiles. Adopted from Kesel (2000).

observed for a free-stream disturbance intensity of around 0.10%. The hysteresis loop, however, disappears as the free-stream turbulence intensity is increased to 0.30%. They suggested that the surface roughness can also produce the same result. Furthermore, the disappearance of the hysteresis loop for aerodynamic lift and drag coefficients at high free-stream turbulence intensity seems to be related to the change in flow structure.

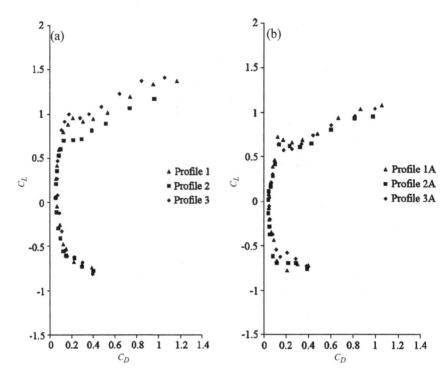

Figure 2.20. Comparison of the lift–drag polars at $Re = 10,000$, for (a) three natural and (b) filled profiles. Adopted from Kesel (2000).

2.2.4 *Effect of Unsteady Free-Stream*

The real operating condition for MAVs is quite different from the conventional low-turbulence wind-/water-tunnel setup. In real flight, MAVs often operate in gusty environments. The effect of unsteady flow on transition was studied by Obremski and Fejer (1967). They experimented with a flat plate subject to a free-stream velocity varying sinusoidally with a mean:

$$U = U_{\text{ref}}\left[1 + N_A \sin(\omega t)\right], \tag{2.21}$$

where N_A is the amplitude ratio, ω is the frequency, and the reference velocity U_{ref} is the mean free-stream velocity. They found that the transition Reynolds number is affected by the free-stream oscillation when the so-called "nonsteady Reynolds number," $Re_{\text{ns}} = N_A U_{\text{ref}}^2/\omega\upsilon$, is above a critical point of about 2.6×10^4. Below the critical value, the unsteady free-stream has little impact on the transition process. Obremski and Morkovin (1969) observed that, in both high and low Re_{ns} ranges, the initial turbulent bursts are preceded in space and time by a disturbance wave packet. By applying a quasi-steady stability model, they concluded that in the high Re_{ns} range the wave packet is amplified rapidly and bursts into turbulence, whereas in the low range the wave packet bursts into turbulence at a much higher Reynolds number. Guided by their study, Lian and Shyy (2006) investigated the influence of free-stream oscillations on the transition for separated flows. In their first test, they

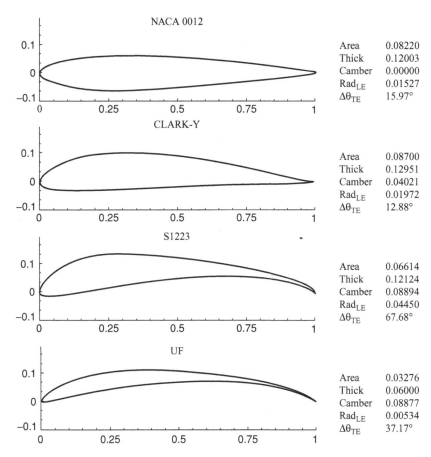

Figure 2.21. Four airfoils chosen for assessment (Shyy et al., 1999b).

set $N_A = 0.33$ and $\omega = 0.3$, resulting in a Strouhal number of 0.0318 and a nonsteady Reynolds number of 9.9×10^4. The frequency ω is kept well below the range of the expected unstable TS wave frequency, which is around 10 Hz.

Figure 2.30 shows the lift coefficient and the lift-to-drag ratio during one selected cycle. Clearly, in a gust situation, the aerodynamic parameters display hysteresis. For example, when flow accelerates (the Reynolds number increases from 6×10^4 to 8×10^4), the lift coefficient does not immediately reach its corresponding steady-state value. Instead, the steady-state value is reached in the decelerating stage. Compared with a steady incoming flow, the gust leads to a higher lift coefficient at the low-velocity end and a lower lift coefficient at the high-velocity end. The lift-to-drag ratio variation during one cycle is substantial. For example, at a Reynolds number of 6×10^4, the lift-to-drag ratio with a steady-state free-stream is around 26; for gust flow, the instantaneous lift-to-drag ratio reduces to 20 when the flow accelerates, but elevates to 38 when the flow decelerates.

Along with the variations in lift and drag, the transition position is also affected by the gust. As shown in Figure 2.31, the transition position moves toward the leading edge when the flow is accelerating and moves toward the trailing edge when flow is

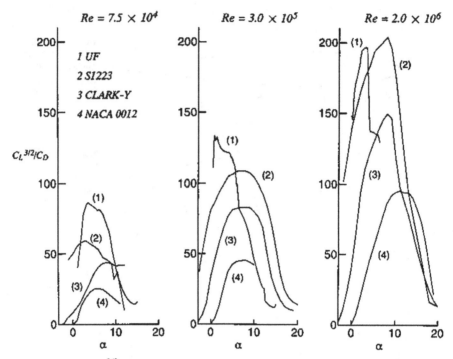

Figure 2.22. $C_L^{3/2}/C_D$ plots at three different Reynolds numbers (Shyy et al., 1999b).

decelerating. During the accelerating stage, the instantaneous Reynolds number is increasing. As the Reynolds number increases, flow experiences early transition. In the simulation of Lian and Shyy (2006), the transition point is simply linked to the computational grid point without further smoothing, resulting in the stair-stepped plot in Figure 2.31.

Lian and Shyy (2006) also investigated a higher frequency of $\omega = 1.5$, five times higher than the previous case, resulting in a nonsteady Reynolds number of 1.98×10^4, which is lower than the critical value. Their numerical result shows that the transition position varies with the instantaneous Reynolds number (Figure 2.31). This seemingly contradicts the observation of Obremski and Morkovin (1969). However, it should be noted that Obremski and Morkovin drew the conclusion based on experiment over a flat plate at a high Reynolds number (10^6), in which the flow is the Blasius type and experiences natural transition. In the test of Lian and Shyy, the separated flow amplifies the unstable TS wave at such a great rate that it results in faster transition to turbulence, typical of the bypass-transition process.

Comparison of the transition position at two different nonsteady Reynolds numbers reveals that flow experiences transition for the whole oscillation cycle at a higher nonsteady Reynolds number, whereas at the lower value the flow becomes laminar at the early accelerating state and remains such until the instant Reynolds number reaches around 7×10^4. It is possible that during the decelerating stage the transition position moves toward the trailing edge because of the lowered Reynolds number. At a higher nonsteady Reynolds number, i.e., lower frequency, the deceleration has

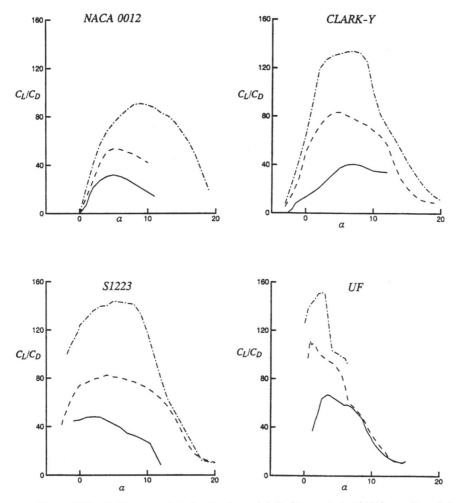

Figure 2.23. C_L/C_D vs. α plots for the four airfoils (Shyy et al., 1999b). —, $Re = 7.5 \times 10^4$; -----, $Re = 3.0 \times 10^5$; ·····, $Re = 2.0 \times 10^6$.

less impact on the transition and the LSB can sustain itself; at a lower nonsteady Reynolds number, i.e., higher frequency, the deceleration has more impact on the transition and the LSB cannot adjust itself with the high rate change to maintain the closed bubble and the LSB bursts. A closed LSB forms only when the Reynolds number reaches 7×10^4. To better appreciate this phenomenon, see the phase and shape factor during one cycle plotted in Figure 2.32.

Another interesting observation at $Re_{ns} = 1.98 \times 10^4$ is the drag coefficient shown in Figure 2.33. During the decelerating stage the gusty flow produces thrust. Analysis shows that the thrust is due to the friction force.

2.3 Three-Dimensional Wing Aerodynamics

Low Reynolds number flyers use low-AR wings, typically no larger than 5. For the MAVs developed by Ifju et al. (2002) the AR is close to 1. Consequently, it is

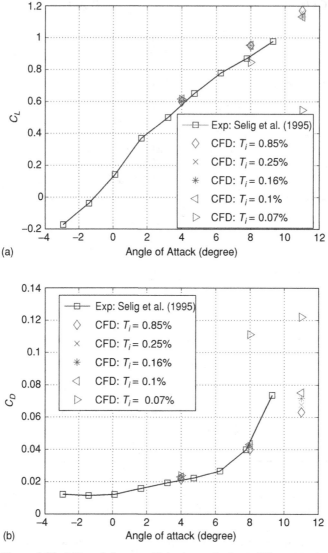

Figure 2.24. Lift and drag coefficients vs. AoA at different turbulence levels for the SD7003 airfoil at $Re = 6 \times 10^4$: (a) lift coefficient, (b) drag coefficient (Lian and Shyy, 2006).

important to investigate the 3D flow structures around a low Reynolds number and low-AR wing.

Lian and Shyy (2005) and Viieru et al. (2005) reported on flow structures around a low-AR rigid wing. The geometry follows the design of Ifju et al., as discussed early in Chapter 2. The wing has a span of 15 cm, a camber of 6%, a root chord of 13.3 cm, and a wing area of 160 cm^2.

To confirm the capabilities of the Navier–Stokes solver, the computational results are first compared with wind-tunnel data measured for a MAV rigid wing with a

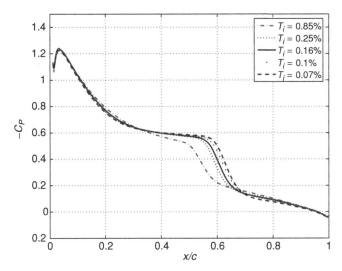

Figure 2.25. Pressure coefficient on the suction surface at $\alpha = 4°$ at different turbulence levels for the SD7003 airfoil at $Re = 6 \times 10^4$ (Lian and Shyy, 2006).

12.5-cm span, which has a smaller area than those used by Lian and Shyy (2005) and Viieru et al. (2005). However, the overall shape and AR are similar.

The experiment is conducted in a horizontal, open-circuit low-speed wind tunnel. It has a square entrance of the bell-mouth-inlet type, and it has several screens that

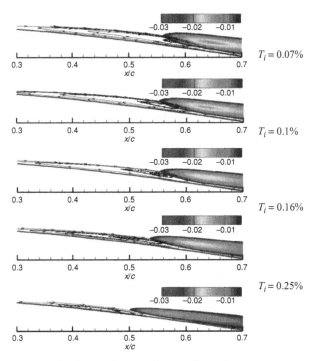

Figure 2.26. Streamlines and normalized shear-stress contours at $\alpha = 4°$ for different turbulence levels for the SD7003 airfoil at $Re = 6 \times 10^4$ (Lian and Shyy, 2006).

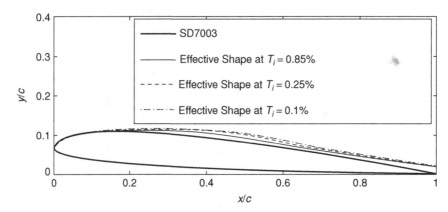

Figure 2.27. Effective airfoil shapes at different turbulence levels for the SD7003 airfoil at $Re = 6 \times 10^4$ (Lian and Shyy, 2006).

provide low turbulence levels, less than 0.1%, in the test section. The test section is 91.4 cm × 91.4 cm and has a length of 2 m. The model under test is attached to a six-component strain-gauge sting balance used to measure the aerodynamic forces and moments. The AoA is controlled by a computer and can be set in any sequence, steady or variable, in time. The force balance is calibrated from 1 to 500 g, from precisely defined loading points. For more detailed information of the experimental measurement and uncertainty, we refer to Albertani et al. (2004).

The 12.5-cm wing configuration is tested at two different Reynolds numbers (7.1×10^4 and 9.1×10^4) based on the root chord length. The experimental data are obtained by averaging of the values from multiple tests for each AoA and Reynolds number. In Figure 2.34(a), the lift versus drag curves are plotted for the two Reynolds numbers just mentioned. The figure demonstrates good agreement between computational and experimental data. As shown in Figure 2.34(b), within the considered Reynolds number range, the lift-to-drag ratio does not vary much. Furthermore, both

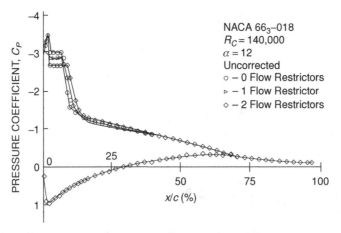

Figure 2.28. Pressure coefficients under different free-stream turbulence levels (O'Meara and Mueller, 1987).

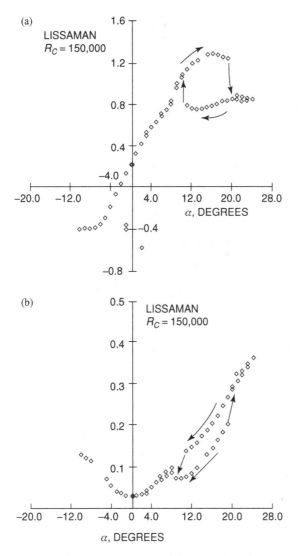

Figure 2.29. Lift and drag coefficients vs. AoA for a smooth Lissaman airfoil (Mueller et al., 1983): (a) lift coefficient, (b) drag coefficient.

experiment and computation show that the best lift-to-drag ratio is reached for an AoA between 4° and 9°.

2.3.1 *Unsteady Phenomena at High Angles of Attack*

Vortex shedding causes more than just unsteadiness in aerodynamic performance. Cummings et al. (2003) reported that, at large AoAs, the unsteady computations predicted noticeably lower lift coefficients than do the steady computations. The Reynolds number in their study is higher than that of the MAV regime. Lian and Shyy (2003) performed Navier–Stokes flow computations around a low-AR wing under MAV flight conditions and found that the differences between the steady-state

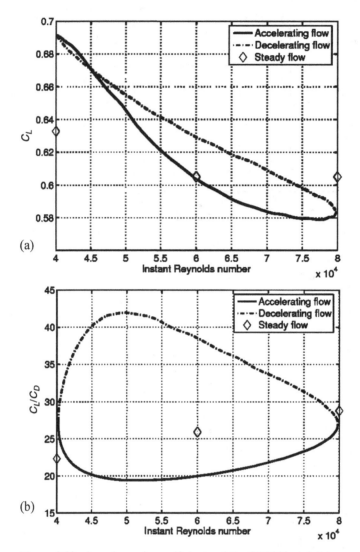

Figure 2.30. Aerodynamic coefficients of the SD7003 airfoil in a gusty environment during one cycle for nonsteady Reynolds number $Re_{ns} = 9.9 \times 10^4$, showing the hysteresis phenomenon: (a) lift coefficient, (b) lift-to-drag ratio (Lian and Shyy, 2006).

and the time-averaged lifts are small even at large AoAs in which unsteady phenomenon such as vortex shedding are prominent. Nevertheless, the instantaneous flow structure varies substantially. Hence it can be misleading to simply examine the time-averaged flow field to estimate the MAV aerodynamic characteristics.

Figure 2.35 compares the pressure coefficients of a MAV wing designed by Ifju and coworkers (2002), which is based on time-averaged unsteady computations and steady-state computations. In this design, the camber gradually decreases from the root toward the tip of the wing. Hence the flow tends to separate first in the root region. At $\alpha = 6°$ the time-averaged pressure coefficient closely matches the steady-state result. The time-averaged value yields a smooth pressure distribution; the

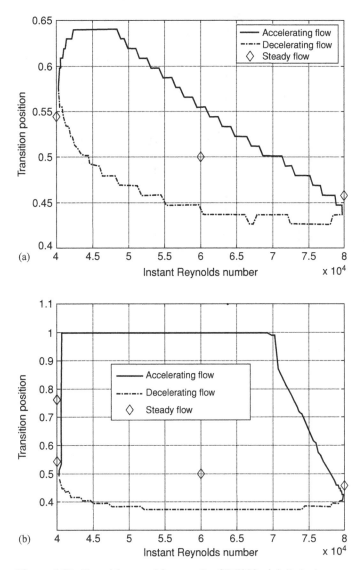

Figure 2.31. Transition position on the SD7003 airfoil during one cycle of the gust: (a) nonsteady Reynolds number $Re_{ns} = 9.9 \times 10^4$, (b) nonsteady Reynolds number $Re_{ns} = 1.98 \times 10^4$ (Lian and Shyy, 2006).

steady-state result indicates a small recirculation zone. As the AoA becomes higher, there is little difference in the leading-edge region; on the contrary, clear differences exist in the separated regions.

2.3.2 *Aspect Ratio and Tip Vortices*

Tip vortices exist on a finite wing because of the pressure difference between the upper and the lower wing surface. The tip vortex establishes a circulatory motion over the wing surface and exerts great influence on the wing aerodynamics.

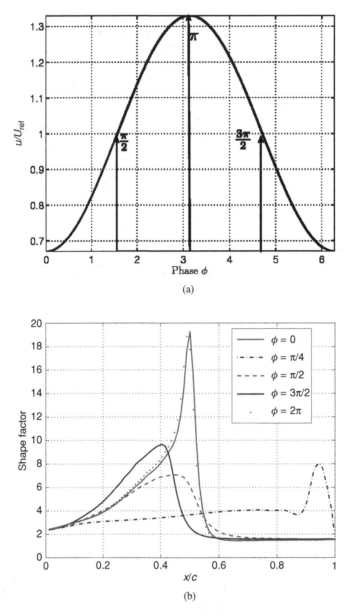

Figure 2.32. Phase and shape factors during one gust cycle on an SD7003 airfoil at the nonsteady Reynolds number $Re_{ns} = 1.98 \times 10^4$: (a) phase, (b) shape factor (Lian and Shyy, 2006).

Specifically, the tip vortex increases the drag force. The total drag coefficient for a finite wing at subsonic speed can be written as (Anderson, 1989)

$$C_D = C_{D,P} + C_{D,F} + \frac{C_L^2}{\pi e \, \text{AR}}, \tag{2.22}$$

where $C_{D,P}$ is the drag coefficient that is due to pressure, $C_{D,F}$ is the drag coefficient that is due to skin friction, e is the span efficiency factor that is less than 1, AR

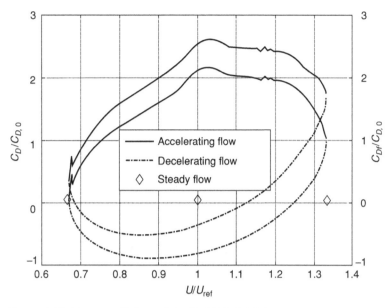

Figure 2.33. Drag coefficient of the SD7003 airfoil in a gusty environment during one cycle for a nonsteady Reynolds number, $Re_{ns} = 1.98 \times 10^4$ (Lian and Shyy, 2006).

is the aspect ratio, and $\frac{C_L^2}{\pi e \mathrm{AR}} = C_{D,i}$ is the induced drag coefficient that is due to the existence of tip vortices. Equation (2.22) demonstrates that the induced drag varies as the square of the lift coefficient; at high AoAs, the induced drag can be a substantial portion of the total drag. Furthermore, Eq. (2.22) illustrates that, as the AR is decreased, the induced drag increases. The MAV wing presented by Ifju et al. (2002) has a low AR of 1.4; therefore it is important to investigate tip vortex effects on the wing aerodynamics. In general, tip vortex effects are twofold:

1. The tip vortex causes downwash that decreases the effective AoA and increases the drag force (Anderson, 1989).
2. The tip vortex forms a low-pressure region on the top surface of the wing, which provides additional lift force (Mueller and DeLaurier, 2003).

Figure 2.36 shows tip vortices around the wing surface together with the streamlines at an angle of attack of 39° (Lian et al., 2003b). The vortical flow is usually associated with a low-pressure zone, as shown in Figure 2.37. The pressure drop further strengthens the swirl by attracting more fluid toward the vortex core; meanwhile, the pressure decreases correspondingly in the vortex core. The low-pressure region created by the vortex generates additional lift. Toward the downstream direction, the pressure recovers to its ambient value, the swirling weakens, the diameter of the vortex core increases, and the vortex core loses its coherent structure.

In Figure 2.38 the evolution of the vortical structure with increasing AoA is visualized. The pressure distribution on the upper surface is also presented in the same figure. At $\alpha = 6°$, tip vortices are clearly visible even though they cover a small area and are of modest strength. The flow is attached to the upper surface and

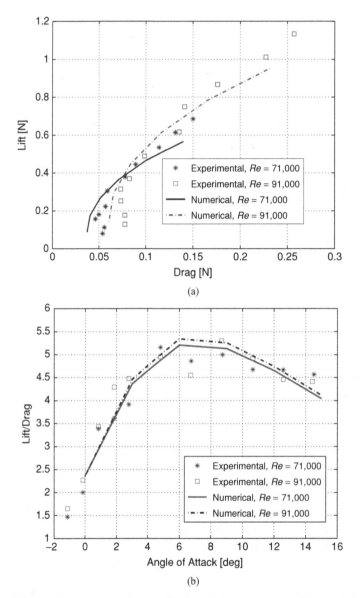

Figure 2.34. Numerical and experimental assessments of lift and drag over a MAV wing for different Reynolds numbers and AoAs (Viieru et al., 2005): (a) polar curve; (b) lift-to-drag ratio vs. the AoA.

follows the chord direction. A low-pressure region near the tip, caused by the vortical structure there, is observed.

Even though the flow on the upper surface near the root tends to separate, the flow remains attached in the outer portion of the wing, and hence the lift still increases with the AoA until, of course, massive separation occurs on most of the upper surface. For low-AR wings, tip vortices make considerable contributions to the lift. This case is similar to that for delta wings. In his numerical study, Lian (2003) observed that the low-AR wing suffered less from separation. The wing is not subjected to sudden

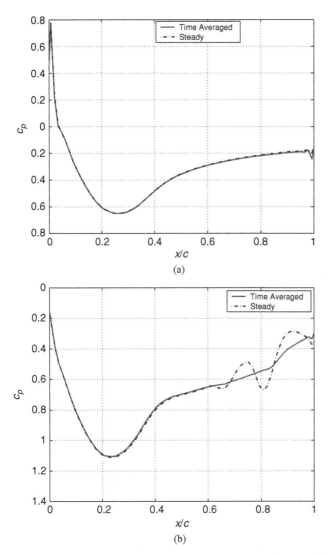

Figure 2.35. Comparisons of c_p on a rigid wing at the root for steady and unsteady computations: (a) $\alpha = 6°$; (b) $\alpha = 15°$, adopted from Lian and Shyy (2003).

stall, but the lift coefficient levels off at very high AoAs. Torres and Mueller (2001), in their experiments on low-AR wings observed similar findings. It should be noted that neither fuselage nor propeller is included in the analysis by Lian (2003).

This pressure drop can be seen from Figure 2.39(a), where the spanwise pressure coefficient on the upper-wing surface at $x/c = 0.4$ is plotted. At $\alpha = 6°$ the spanwise pressure is almost uniform on the upper-wing surface, and the tip vortex causes the pressure drop to occur at approximately 90% of the half-span from the root. Figure 2.39 is illustrative in regard to pressure distributions versus the vortical structures. They are not indicative of the total level of the pressure force.

Vortices strengthen with an increase in the AoA. At $\alpha = 27°$, as shown in Figure 2.38, tip vortices develop a strong swirl motion while entraining the surrounding flow. The low-pressure area increases as the AoA becomes higher. In

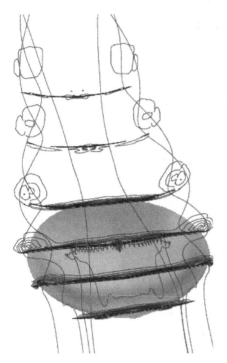

Figure 2.36. Streamlines and vortices for a rigid wing at $\alpha = 39°$. The vortical structures are shown on selected planes (Lian et al. 2003b). (See Plate IX.)

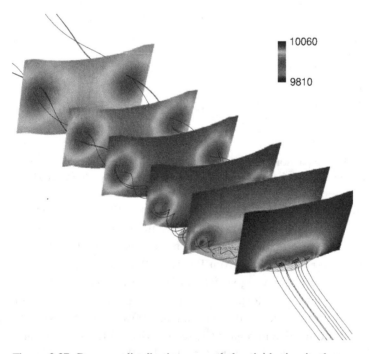

Figure 2.37. Pressure distribution around the rigid wing in the cross sections with streamlines at an AoA of 39° (Lian et al. 2003b). (See Plate X.)

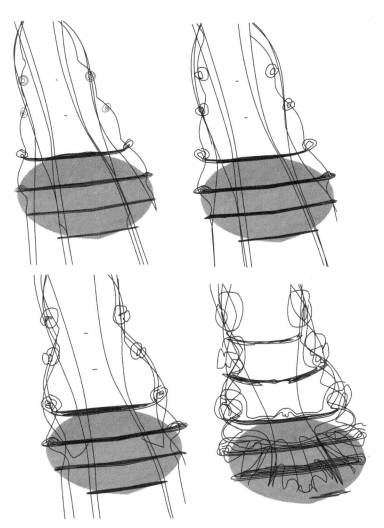

Figure 2.38. Evolution of flow pattern for rigid wing vs. AoAs. From left to right, top to bottom, 6°, 15°, 27°, and 51°, from Lian and Shyy (2005). (See Plate XI.)

Figure 2.39(a), the pressure drop moves along the spanwise direction toward the root and now occurs at 75% from the root.

At lower AoAs, the vortex core position shows a linear relation with the incidence. This relation disappears at higher AoAs when the flow is separated on the upper surface. For example, at $\alpha = 45°$, the flow is separated at the leading edge, and the low-pressure zone covers more than 40% of the wing surface, which helps to maintain the increase in lift force. At $\alpha = 51°$, a considerable spanwise velocity component is seen and the flow is separated from most of the upper surface (Figure 2.38). The separation on the upper-wing surface decreases the lift, and stall occurs.

As observed before, the tip vortices have an important effect on the low-AR wing aerodynamics. One major effect of the tip vortices is the increase in induced drag

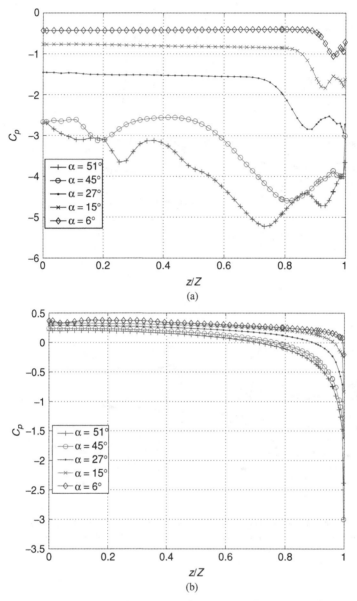

Figure 2.39. Spanwise pressure coefficient distributions at $x/c = 0.4$ for a rigid wing at different AoAs: (a) pressure coefficient at upper surface, (b) pressure coefficient at lower surface. Adopted from Lian and Shyy (2005).

for low-AR wings. Equation (2.22) shows that, the smaller the AR, the larger the induced drag.

2.3.3 Wingtip Effect

The wing shape chosen here strives to maximize the wing area, and hence the lift, for a given dimension. However, the tip vortices associated with the present low-AR wing also substantially affect the aerodynamics. It is well established that

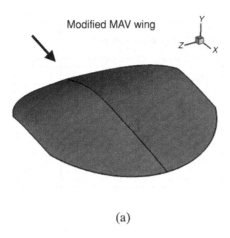

(a)

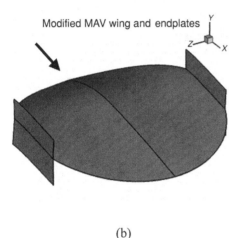

(b)

Figure 2.40. Wing-shape geometry: (a) modified wing, (b) endplates' location on the modified wing (Viieru et al., 2005).

the tip vortex causes a downwash that modifies the pressure distribution on the wing surface and increases the induced drag. Various methods to reduce the induced drag by decreasing the tip vortex effects are described in the literature and confirmed by actual applications to aircraft wing design (LaRoche and Palffy, 1996). Viieru et al. (2005) reported the implication of placing endplates at the wingtip, which is simple from the manufacturing point of view.

The effect of endplates on the MAV rigid-wing aerodynamics was previously investigated by Viieru et al. (2003). In that study the endplate was simply added to the existing MAV wing to probe its effect on the tip vortex and overall aerodynamics; the wing shape was retained. It was observed that the endplate increases lift by reducing the downwash and increases the effective AoA. However, drag increases along with the curved endplate in part because the endplate behaves as a vertically placed airfoil, and the additional form drag causes the overall lift-to-drag ratio to decrease.

To remedy the disadvantages of the endplates, Viieru et al. (2005) investigated alternative configurations. Three wing geometries were studied: the original wing

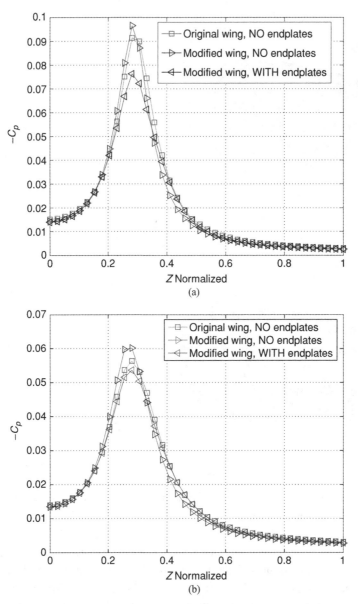

Figure 2.41. Pressure coefficient along the vortex core behind the wing at $6°$ AoA: (a) $x/c = 3$; (b) $x/c = 5$ (Viieru et al., 2005).

discussed at the beginning of this chapter, a modified wing (Figure 2.40(a)) with trimmed tip, and a modified wing with endplates (Figure 2.40(b)). Compared with the original wing, the trimmed wing has a shorter span of 14 cm and a small wing area of 155 cm^2, whereas the root chord has the same length as the original wing. The endplate attached to the modified wing, which is parallel to the flight direction, has a length of 4.4 cm and a height of 3.4 cm.

One can observe the vortex intensity and the circulation by looking at the slices perpendicular to the streamwise direction behind the wing. Behind the trailing edge, the flow can be approximated with a vortex core of constant rotation and a potential motion outside the core. The relation between the pressure at the vortex center and the circulation around a rigid rotating body is given by (Prandtl and Tietjens, 1957)

$$\Gamma^2 = 4\pi^2 r_1^2 p_{center}/\rho, \tag{2.23}$$

where r_1 is the rigid-body radius, p_{center} is the pressure at the rigid-body center, and ρ is the fluid density. Equation (2.23) shows that the vortex strength, measured by its circulation, is proportional to the pressure drop in the vortex core and its radius. In Figure 2.41, the pressure coefficient is plotted along the vortex core diameter at $x/c = 3$ behind the wing and $x/c = 5$. The amount of pressure drop inside the vortex core indicates that the endplates reduce the vortex strength. Also, the modified wing without the endplates shows the strongest vortex.

From the pressure contours and horizontal velocity contours, one observes that the endplate affects the flow field over the wing. The endplate slows down the flow near the wingtip. This decrease in velocity reduces the pressure drop on the upper-wing surface that corresponds to the vortex core (Figure 2.42(a)). On the other hand, a lower velocity slightly below the wing increases the high-pressure area there because more momentum is transferred to the wing as pressure instead of being shed as vorticity at the wing tip. The increase in the high-pressure zone on the lower-wing surface in the presence of the endplate can be clearly seen from the spanwise pressure coefficient on the lower-wing surface plot (Figure 2.42(b)).

In Figure 2.43 the spanwise lift distribution obtained by integration of the pressure of difference along the local chord at a specified spanwise location is plotted. It clearly shows that, when the endplates are attached, the lift on each cross section is higher compared with that of the wing without the endplate. With a smaller overall wing area, the modified wing with the endplates produces almost the same lift as the original wing. Furthermore, the modified wings (with and without endplate) experience lower drag over almost 75% of the wingspan, starting from the root.

In Table 2.1 the overall aerodynamic performances parameters are presented for a 6° AoA. The modified wing configuration with endplates has a better lift-to-drag ratio than the baseline configuration (10% improvement). This improvement is mainly due to the drag reduction by the modified wing shape because the total lift is essentially the same. In Table 2.2 the same parameters are presented for an AoA of 15°. The modified wing with endplates shows an increase of 1.4% in the lift-to-drag ratio compared with the baseline configuration.

2.3.4 Unsteady Tip Vortices

A wing with a low AR is susceptible to rolling instabilities (wobbling). This concern is particularly important in view of the strong gust effect on MAVs. Tang and Zhu (2004) investigated the aerodynamic characteristic of a low-AR wing. The wing has an elliptic planform, using the E174 airfoil with an AR ratio of 1.33. Based on the maximum chord length, the Reynolds number is 10,000. Through numerical simulation and flow visualization in a water tunnel, they found that tip vortices are

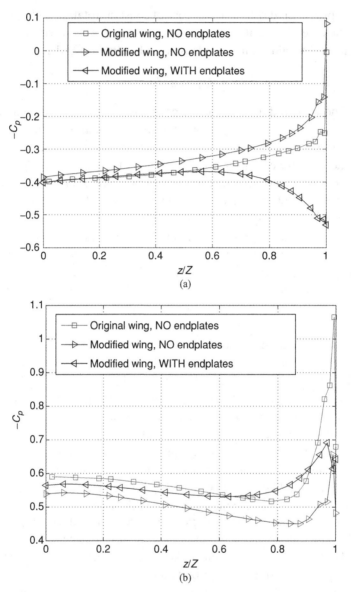

Figure 2.42. Pressure coefficient on the wing surface at $x/c = 0.34$ and 6° AoA: (a) lower-wing surface; (b) upper-wing surface (Viieru et al., 2005).

unsteady in sizes and strengths when the AoA is larger than 11°. Figures 2.44(a)–2.44(c), on the right-hand side, show positions of the tip vortices at an AoA of 25° in the vertical plane (Trefftz plane) at three time instants. As time evolves, the left-hand and right-hand tip vortices change their sizes and strengths. The asymmetric flow causes unequal drag between the two sides of the wing, which produces a yawing moment; the asymmetric flow also causes uneven lift, resulting in a rolling instability.

From the numerical results, they suggested that this unstable phenomenon is caused by the interaction between the secondary vortical flows and the tip vortices.

Table 2.1. *Aerodynamic forces at a 6° AoA (Viieru et al., 2005)*

AoA = 6°	Original MAV wing, no endplates	Modified MAV wing, no endplates	Modified MAV wing, with endplates
Lift (N)	0.49	0.44	0.49
Drag (N)	0.074	0.065	0.067
Lift / drag	6.64	6.85	7.39

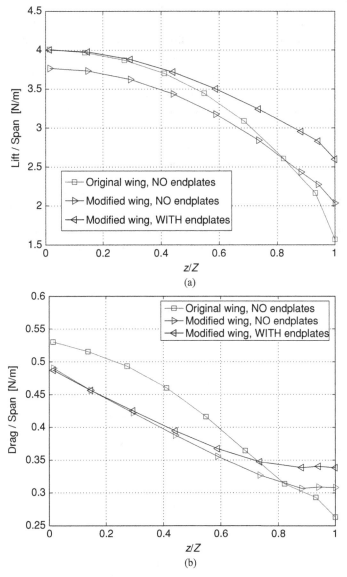

Figure 2.43. Spanwise lift and drag distribution at 6° AoA: (a) lift, (b) drag (Viieru et al., 2005).

Table 2.2. *Aerodynamic forces at a 15° AoA (Viieru et al., 2005)*

AoA = 15°	Original MAV wing, no endplates	Modified MAV wing, no endplates	Modified MAV wing, with endplates
Lift (N)	0.92	0.86	0.87
Drag (N)	0.22	0.21	0.21
Lift / drag	4.16	4.15	4.22

The separated vortical flows are on the upper surface of the wing. The schematic figure on the left-hand side of Figure 2.44 shows that, as the wing incidence progressively increases from 5°, substantial time dependency of the tip vortices is observed. At $\alpha = 5°$, the position of the separated vortical flow is around the trailing edge. As the incidence increases, the separating flow moves toward the leading edge. When the incidence reaches 15° or higher, the separating flows above the wing interact with the tip vortices, causing the tip vortices to become substantially unsteady. To date, the MAV flight test has not reported such rolling instabilities as a major barrier. This is apparently because the airfoil shapes used for MAV flyers are much thinner and do not induce as many separating flows above the wing surface. Nevertheless, the issue of unsteady tip vortices needs be investigated in the MAV design and flight test process.

2.4   Concluding Remarks

In this chapter we presented the fixed, rigid-wing aerodynamics at low Reynolds number ranges, between 10^3 and 10^6. The main points are summarized as follows:

1. The maximum lift-to-drag ratio of an airfoil decreases substantially as the Reynolds number drops from 10^6 to 10^4 or lower.

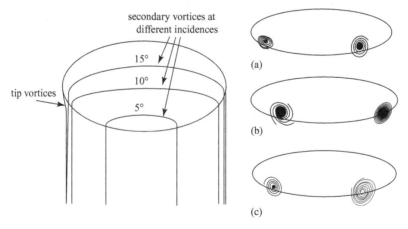

Figure 2.44. Left: Schematic of the dynamics of tip vortices (viewed above the wing, secondary vortices are above the upper surface of the wing); right: tip vortex stream-lines in vertical planes (Trefftz plane) at about $0.5c$ behind the trailing edge at an AoA of 25°, at three nondimensional times (based on the free-stream velocity and maximum chord length): (a) $t = 42$, (b) $t = 54$, and (c) $t = 62$ (viewed from aft). Adopted from Tang and Zhu (2004).

2. Overall, a thinner airfoil with modest camber is preferable for low Reynolds number flyers because it generates better lift-to-drag ratio and better power efficiency.

3. The laminar-to-turbulent transition and LSB play important roles in determining the airfoil performance at a Reynolds number around 10^4. In this flow regime, the lift-to-drag polar exhibits zigzag characteristics that are due to the formation and burst of the LSB. Because of the effect of transition, the wing performance is expected to be sensitive to the free-stream turbulence intensity and gust.

4. In the Reynolds number range between 10^3 and 10^4, a corrugated wing can provide a more favorable lift than a noncorrugated wing because the viscous effect substantially modifies the effective airfoil shape.

5. Wind gust is a prominent factor in low Reynolds number flyers. Low Reynolds number aerodynamics often exhibits hysteresis in a gusty environment. The transition position varies with the instant Reynolds number, and, it should be noted that, depending on the flow parameters, either drag or thrust can be generated from the unsteady aerodynamics.

6. Tip vortex induces a downwash movement, which reduces the effective AoA of a wing. For a low-AR, low Reynolds number wing, the induced drag by the tip vortex substantially affects its aerodynamic performance. They not only affect lift and drag generation, but also potentially flight stability.

CHAPTER THREE

Flexible-Wing Aerodynamics

3.1   General Background of Flexible-Wing Flyers

In the development of MAVs, there are three main approaches, which are based on flapping-wings, rotating wings, and fixed wings for generating lift. We focus on the fixed, flexible-wing aerodynamics in this chapter. It is well known that flying animals typically have flexible wings to adapt to the flow environment. Birds have different layers of feathers, all flexible and often connected to each other. Hence, they can adjust the wing planform for a particular flight mode. The flapping modes of bats are more complicated than those of birds. Bats have more than two dozen independently controlled joints in the wing (Swartz, 1997) and highly deforming bones (Swartz et al., 1992) that enable them to fly at either a positive or a negative AoA, to dynamically change wing camber, and to create a complex 3D wing topology to achieve extraordinary flight performance. Bats have compliant thin-membrane surfaces, and their flight is characterized by highly unsteady and 3D wing motions (Figure 3.1). Measurements by Tian et al. (2006) have shown that bats exhibit highly articulated motion, in complete contrast to the relatively simple flapping motion of birds and insects. They have shown that bats can execute a 180° turn in a compact and fast manner: flying in and turning back in the space of less than one half of its wing-span and accomplishing the turn within three wing beats with turn rates exceeding 200°/s.

Birds and bats can also change the span (flexing their wings) to decrease the wing area, increase the forward velocity, or reduce drag during an upstroke. In fast forward flight, birds and bats reduce their wing area slightly during the upstroke relative to the downstroke. At intermediate flight speeds, the flexion during the upstroke becomes more pronounced. However, bats and birds flex their wings in different manners. The wing-surface area of a bird's wing consists mostly of feathers, which can slide over each other as the wing is flexed and still maintain a smooth surface. Bat wings, in contrast, are mostly thin membrane supported by the arm bones and the enormously elongated finger bones. Given the stretchiness of the wing membrane, bats can flex their wings a little, reducing the span by about 20%, but they cannot flex their wings too much or the wing membrane will go slack. Slack membranes are inefficient, because drag goes up, and the trailing edges are prone to flutter, making them more difficult for fast flight (Alexander, 2002).

Plate I. Maneuvering capabilities of natural flyers. (a) Canada geese's response to wind gust; (b) speed control and target tracking of a seagull; (c) precision touch-down of a finch; (d) a hummingbird defending itself against a bee.

Plate II. Natural flyers can track target precisely and instantaneously. Shown here are hummingbirds using flapping wings, contoured body, and tail adjustment to conduct flight control.

Plates I-XXIX are available for download in color from www.cambridge.org/9780521204019

Plate III. Natural flyers synchorize wings, body, legs, and tail to take off, on water (top), from land (middle), and off tree (bottom).

Plate IV. Birds such as seagulls glide while flexing their wings to adjust their speed as well as to control their direction.

Plate V. On landing, birds fold their wings to reduce lift, and flap to accommodate wind gusts and to adjust for their available landing areas.

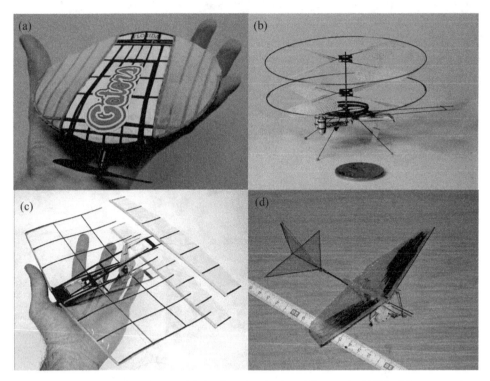

Plate VI. Representative MAVs. (a) flexible fixed wing (Ifju et al., 2002); (b) rotary wing (http://www.proxflyer.com); (c) hybrid flapping-fixed wing, using fixed wing for lift and flapping wing for thrust (Jones and Platzer, 2006); and (d) flapping wing for both lift and thrust (Kawamura et al., 2006).

Plate VII. Illustration of biological flapping-wing patterns: forward and back strokes, and flexible- and asymmetric-wing motions.

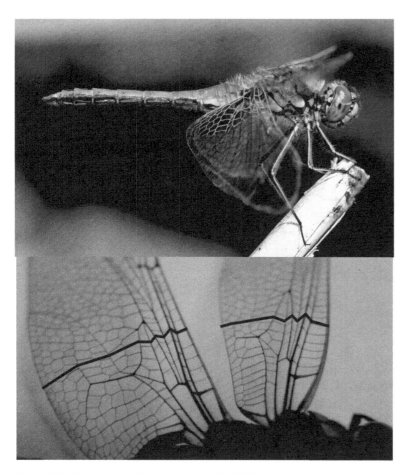

Plate VIII. Dragonfly wings exbit both flexibility and anisotropic, corrugated structuures. In the lower picture, shown on the left is the hind wing and the right is the fore wing.

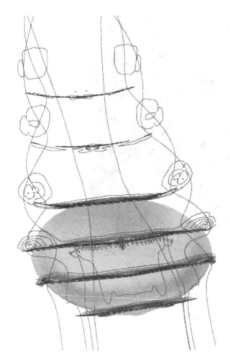

Plate IX. Streamlines and vortices for rigid wing at $\alpha = 39°$. The vortical structures are shown on selected planes (Lian et al., 2003b).

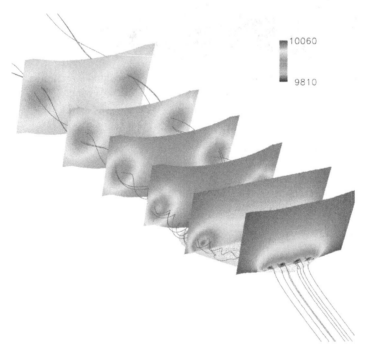

Plate X. Pressure distribution around the rigid wing in the cross sections with streamlines at angle of attack of 39° (Lian et al., 2003b).

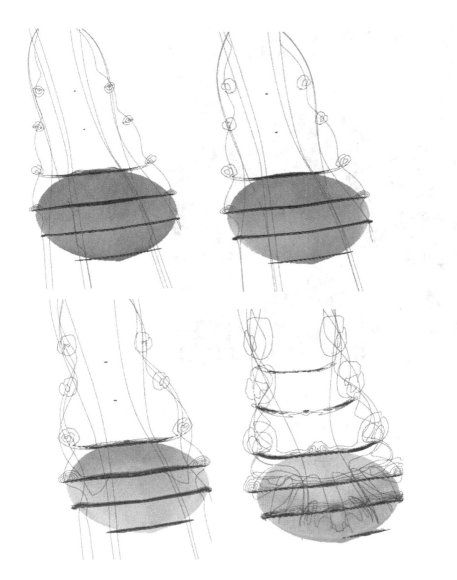

Plate XI. Evolution of flow pattern for rigid wing versus angles of attack. (From left to right, top to bottom, 6°, 15°, 27°, and 51°) (Lian and Shyy, 2005).

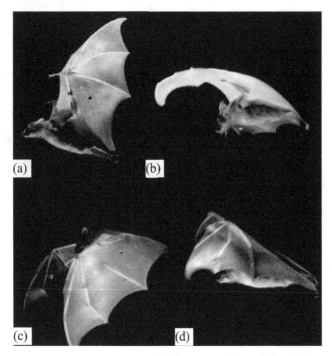

Plate XII. A bat *(Cynopterus brachyotis)* in flight. (a) beginning of downstroke, head forward, tail backward, the whole body is stretched and lined up in a straight line; (b) middle of downstroke, the wing is highly cambered; (c) end of downstorke (also beginning of upstroke), the wing is still cambered. A large part of the wing is in front of the head and the wing is going to be withdrawn to its body; (d) Middle of upstroke, the wing is folded towards the body, from Tian et al. (2006).

flap

Plate XIII. The flexible covert feathers acting like self-activated flaps on the upper wing surface of a skua. Photo from Bechert et al. (1997).

Plate XIV. Vortices structure behind a stationary NACA 0012 (Lai and Platzer, 1999).

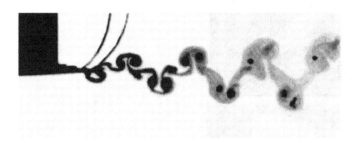

(a) $h = 0.0125$ ($kh = 0.098$)

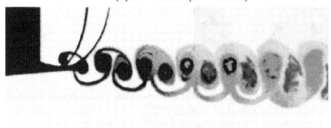

(b) $h = 0.025$ ($kh = 0.196$)

(c) $h = 0.05$ ($kh = 0.393$)

Plate XV. Vortex patterns for a NACA 0012 airfoil oscillated in plunge for a freestream velocity of about 0.2 m/s, a frequency of $f = 2.5$ Hz ($k = 7.85$), and various amplitudes of oscillation (Lai and Platzer, 1999).

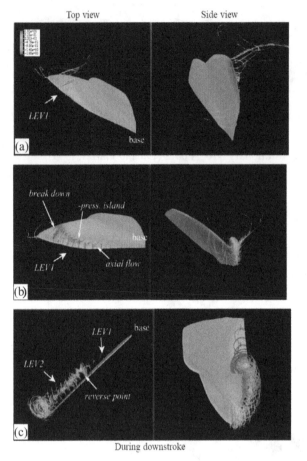

Plate XVI. Wing surface pressure and streamlines revealing the vortical structures for the 3D numerical simulation of a hovering hawkmoth (Liu et al., 1998). (a) Positional angle φ=30°; (b) φ=0°; (c) φ=-36°. Reynolds number is approx. 4000 and the reduced frequency *k* is 0.37. Here LEV is the leading edge vortex.

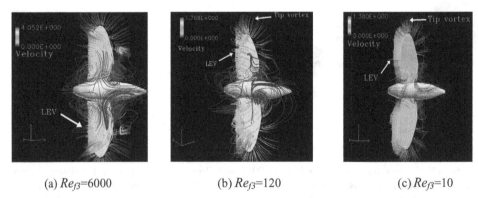

Plate XVII. Numerical results of leading edge vortical structures at different Reynolds numbers.

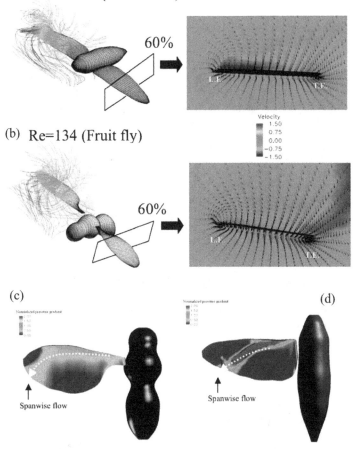

(a) Re=6000 (Hawkmoth)

60%

(b) Re=134 (Fruit fly)

60%

Velocity
1.50
0.75
0.00
-0.75
-1.50

(c)

Normalized pressure gradient

Spanwise flow

(d)

Normalized pressure gradient

Spanwise flow

Plate XVIII. Comparison of near-field flow fields between a fruit fly and a hawkmoth. Wing-body computational model of (a) a hawkmoth ($Re_{f3} = 6000$, $U_{ref} = 5.05$ m/s, $c_m = 1.83$ cm), and (b) a fruit fly model ($Re_{f3} = 134$, $U_{ref} = 2.54$ m/s, $c_m = 0.78$ mm), with the LEVs visualized by instantaneous streamlines and the corresponding velocity vectors in a plane cutting through the left wing at 60% of the wing length; pressure gradient contours on the wing surface for (c) a fruit fly, and (d) a hawkmoth. The pressure gradient indicates the direction of the spanwise flow.

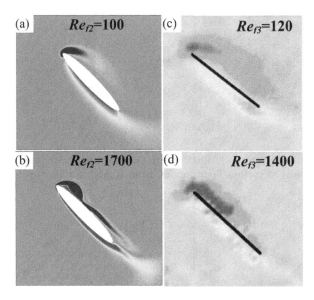

(a) $Re_{f2}=100$

(c) $Re_{f3}=120$

(b) $Re_{f2}=1700$

(d) $Re_{f3}=1400$

Plate XIX. Vortical flow structures for pitch-up airfoils: (a) and (b) computational results for flow over a 2D elliptic airfoil undergoing "water treading" hovering at two Reynolds numbers. The airfoil position corresponds to the mid-stroke, where the pitch angle reaches the maximum value; (c) and (d) experimental vorticity field side views for a fruit fly modeled wing at 0.65R at mid-stroke. The experimental information in (c) and (d) is reprinted from Birch et al. (2004).

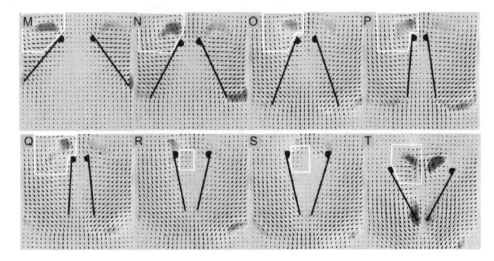

Plate XX. Experiment of clap-and-fling by two wings (M–T) using clap-and-fling wing beat pattern in the robotic wing. Vorticity is plotted according to the pseudo color code and arrows indicate the magnitude of fluid velocity; longer arrows signifying larger velocities, from Lehmann et al. (2005) with permission.

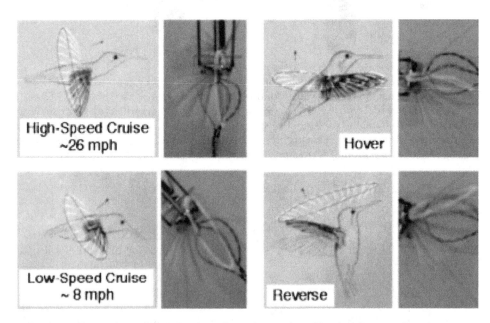

Plate XXI. Comparison of the wingtip trajectories produced by the vibratory flapping system with those exhibited by hummingbirds in various flight modes, from Raney and Slominski (2004).

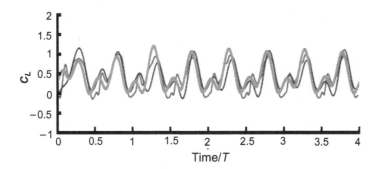

Plate XXII. Numerical and experimental results of the flapping motion of a fruit fly: red, experimental results of Dickinson and Birch (Wang et al., 2004); Blue, numerical solution of Wang et al. (2004); green, numerical solution of Tang et al. (2007). $h_a/c = 1.4$, $\alpha_a = 45°$, $Re_{f2} = 75$, $k = 0.357$.

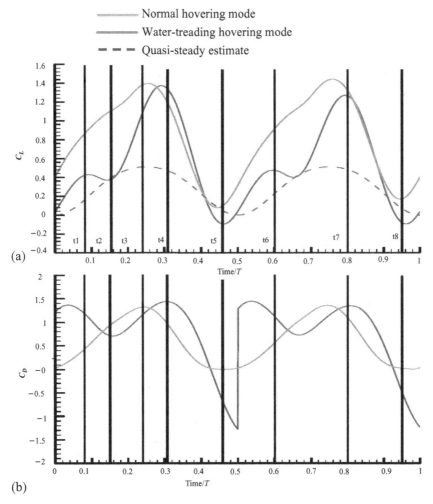

Plate XXIII. One cycle force history for two hovering modes and quasi-steady value of normal hovering mode. $h_a/c = 1.4$, $\alpha_a = 45°$, $k = 0.357$, and $Re_{f2} = 100$. (a) Lift coefficient, (b) drag coefficient. The selected normalized time instants are t1 = 0.08, t2 = 0.17, t3 = 0.25, t4 = 0.31, t5 = 0.45, t6 = 0.60, t7 = 0.80, t8 = 0.94.

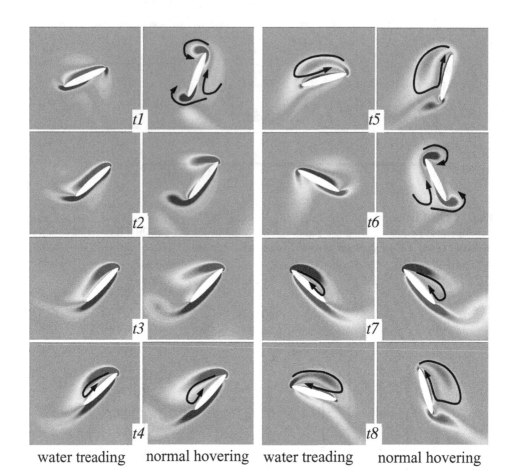

| water treading | normal hovering | water treading | normal hovering |

Plate XXIV. Vorticity contours for two hovering modes. $h_a/c = 1.4$, $\alpha_a = 45°$, $k = 0.357$ and $Re_{f2} = 100$. Red: counter-clockwise vortices, Blue: clockwise vortices. The flow snapshots (t1 to t8) correspond to the time instants defined in Figure 4-38. Adopted from Tang et al. (2007).

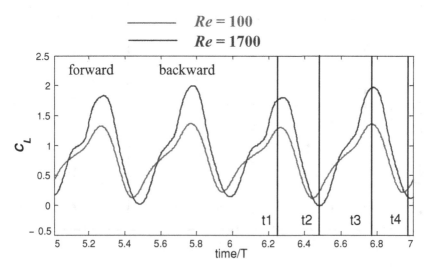

Plate XXV. Lift coefficient for the water-treading mode. $h_a/c = 1.4$, $\alpha_a = 45°$, $k = 0.357$, and Reynolds numbers of 100 and 1700. The selected normalized time instants are t1 = 6.25, t2 = 6.48, t3 = 6.77, t4 = 6.97.

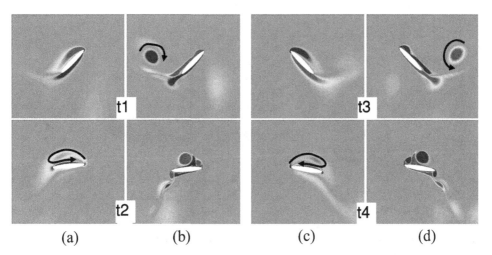

Plate XXVI. Vorticity contours for the "water treading" mode. $h_a/c = 1.4$, $\alpha_a = 45°$, $k = 0.357$. Red = counter-clockwise vortices, Blue = clockwise vortices. (a), (c) Reynolds number $= 100$; (b), (d) Reynolds number $= 1,700$. The flow snapshots (t1 to t4) correspond to the time instants defined in Figure 4.40.

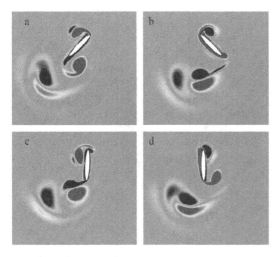

Plate XXVII. Vorticity contours at two corresponding positions during forward (a) and (c) and backward (b) and (d) stroke. Stroke amplitude $h_a/c = 0.25$, pitch angle amplitude $\alpha_a = 45°$ and $Re_{f2} = 300$.

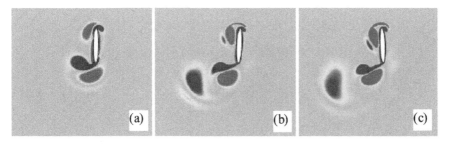

Plate XXVIII. Vorticity contours at time $/T = 5.5$ and three different Reynolds number with a stroke amplitude $h_a/c = 0.25$ and $\alpha_a = 45°$: (a) $Re_{f2} = 75$; (b) $Re_{f2} = 300$; (c) $Re_{f2} = 500$.

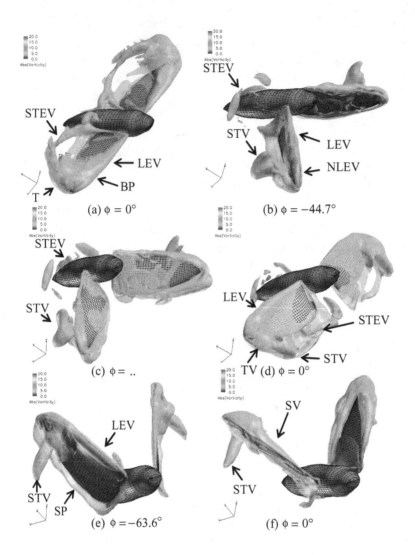

Plate XXIX. Iso-vorticity surfaces (absolute vorticity strengths: 4=green, 13=blue) around flapping wings and body of a hawkmoth during a flapping cycle. Shedding TV (STV) shedding TEV (STEV), new LEV (NLEV), stopping-vortex (SPV), starting-vortex (SV), and break-down point.

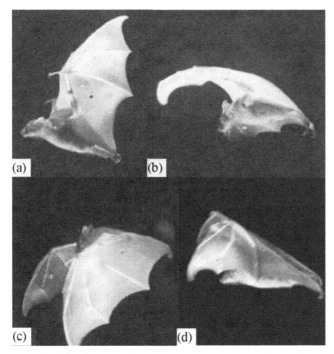

Figure 3.1. A bat (*Cynopterus brachyotis*) in flight: (a) beginning of downstroke, head forward, tail backward, the whole body is stretched and lined up in a straight line; (b) middle of downstroke, the wing is highly cambered; (c) end of downstroke (also beginning of upstroke), the wing is still cambered. A large part of the wing is in front of the head and the wing is going to be withdrawn to its body; (d) middle of upstroke, the wing is folded toward the body, from Tian et al. (2006). (See Plate XII.)

While making bending or twisting movements, biological flyers have natural capabilities of adjusting the camber of their wings in accordance with what the flow environment dictates, such as a wind gust, object avoidance, and target tracking. Bats are known for being able to change the shape of the wing passively, depending on the free-stream conditions. As shown in Figure 3.1, bats can change their wing shapes during each flapping cycle. In manmade devices, sails and parachutes operate under similar ideas. This passive control of the wing surface can prevent flow separation and enhance lift-to-drag ratio. Birds adjust their wings based on different strategies. For example, some species have coverts that act as *self-activated flaps* to prevent flow separation. These features offer shape adaptation and help adjust the aerodynamic control surfaces; they can be especially helpful during landing and in an unsteady environment. In Figure 3.2 the coverts have popped up on a skua and the flexible structure of the feathers is clearly shown.

Nature's design of flexible-membrane wings can be put into practice for MAVs. When a flexible-wing design is adopted (Figures 3.3 and 3.4), similar to that of bat wings, the performance of the MAVs can be improved, especially at high AoAs, by passive shape adaptation, which results in delayed stall (Shyy et al., 1999a; Waszak et al., 2001).

Figure 3.2. The flexible covert feathers acting as self-activated flaps on the upper-wing surface of a skua. Photo from Bechert et al. (1997). (See Plate XIII.)

Figure 3.5, adopted from Waszak et al. (2001), compares the lift curves versus AoAs for rigid and membrane wings. The three different flexible-wing arrangements are depicted in Figure 3.4. The one-batten design has the most flexibility, characterized by large membrane stretch. The two-batten design is, by comparison, stiffer and exhibits less membrane stretch under aerodynamic load. The six-batten wing is covered with an inextensible plastic membrane that further increases the stiffness

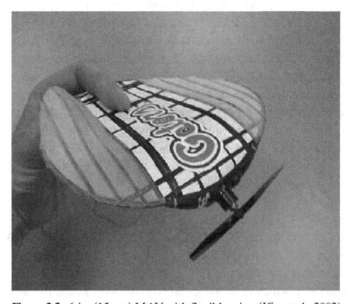

Figure 3.3. 6-in. (15-cm) MAV with flexible wing (Ifju et al., 2002).

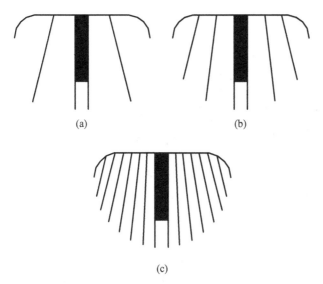

Figure 3.4. Three versions of the flexible wing were tested in the wind tunnel; adopted from Waszak et al. (2001): (a) one-batten flexible wing, (b) two-batten flexible wing, (c) six-batten flexible wing covered with a plastic inextensible membrane.

of the wing and exhibits less membrane deformation and vibration. The nominally rigid wing is constructed of a two-batten frame covered with a rigid graphite sheet.

Under modest AoAs, both rigid and membrane wings demonstrate similar lift characteristics, with the stiffer wings having a slightly higher lift coefficient. However, a membrane wing stalls at substantially higher AoAs than a rigid wing. This aspect is a key element in enhancing the stability and agility of MAVs.

The membrane concept has been successfully incorporated in MAVs designed by Ifju et al. (2002). To implement the flexible-wing concept on these small vehicles, traditional materials such as balsa wood, foam, and monocoat are not appropriate. In their design, illustrated in Figures 3.3 and 3.6, unidirectional carbon fiber and cloth prepreg materials were used for the skeleton (leading-edge spar and chordwise

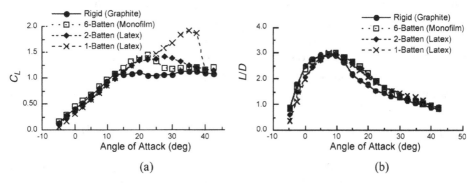

Figure 3.5. Aerodynamic parameters vs. AoA for configurations with varying wing stiffnesses: (a) lift coefficient vs. AoA, (b) lift-to-drag ratio vs. AoA. Adopted from Waszak et al. (2001).

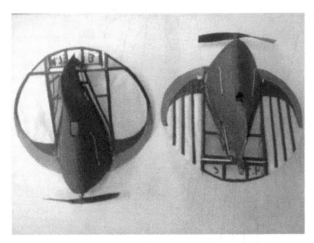

Figure 3.6. Representative MAVs with membrane wing developed by Peter Ifju and collaborators at the University of Florida. Left, the wing is framed along the entire peripherals; right, the wing is flexible along the trailing edge while reinforced by battens.

battens). These are the same materials used for structures that require fully elastic behavior yet undergo large deflections. The fishing rod is a classic example of such a structure. For the membrane, extensible material was chosen to allow deformation even under very small loads, such as the case for lightly loaded wings. Latex rubber sheet material was used in this case. The stiffness of the whole structure can be controlled by the number of battens and membrane material.

As presented earlier, the experimental data for rigid and flexible wings (Figure 3.4), with configurations similar to those shown in Figure 3.6, show that a membrane wing stalls at substantially higher AoAs than a rigid wing (Figure 3.5). Some aspects of low-AR, low Reynolds number rigid-wing aerodynamics were presented by Torres and Mueller (2001). The lift curve slope in Figure 3.5 is approximately 2.9 with the prop pinned. The lift curve slopes of similar rigid wings reported by Torres and Mueller (2001) at comparable Reynolds number and AR ($Re = 7 \times 10^4$, AR = 2) are approximately 2.9 as well. However, these wings have stall angles between 12° and 15°. The stall angles of the flexible wings are between 30° and 45° and are similar to those of much lower-AR rigid wings (AR = 0.5–1.0) (Mueller and DeLaurier, 2003). However, low-AR rigid wings exhibit noticeably lower lift curve slopes, typically between 1.3 and 1.7 (Mueller and DeLaurier, 2003). Hence flexible wings can effectively maintain the desirable lift characteristics with better stall margins (Waszak et al., 2001). Figure 3.6 shows fixed, flexible-wing MAVs designed by Ifju and coworkers of the University of Florida (Ifju et al., 2002). The general specifications of the design are presented in Table 3.1.

From an engineering point of view, flexibility can be used for purposes other than flight quality improvement. These include shape manipulation and reconfiguration for both improved maneuvering and storage. Traditional control surfaces such as rudders, elevators, and ailerons have been used almost exclusively for flight control.

Table 3.1. *General specifications for the UF MAV*

Wing span	4.5 in. (11.43 cm)
Fuselage length	4.5 in. (11.43 cm)
Take-off weight	45 g
Engine	Maxon Re10
Propeller	U-80 (62 mm)
RC receiver	PENTA with customized half-wave antenna
Maximum mission radius	0.9 miles
Video transmitter	SDX-22 70 mw
Camera	Complementary metal-oxide semiconductor camera (350-line resolution)

When the wing is morphed or reshaped by distributed actuation such as piezoelectric and shape-memory material, preferred wing shapes can be developed for specific flight regimes. Such a reconfiguration, however, would require substantial authority and power if the wings were nominally rigid. The flexible nature of the wing allows for such distributed actuation with orders of magnitude less authority. For example, the individual battens on the wing can be made from shape-memory alloys or piezoelectric materials, or traditional actuators, such as servos, which can be used to manipulate the shape and properties of the wing.

Figure 3.7 illustrates a model with morphing technology. It uses a thread connecting the wingtips to a servo in the fuselage of the airplane. As the thread is tightened on one side of the aircraft, it acts as an aileron and causes the AoA of the wing to increase. The roll rate developed by such an actuation mechanism is considerably higher than that from a rudder. Additionally, it produces nearly pure roll with little yaw interaction. Detailed information of the related technical approaches was given by Garcia et al. (2003).

In some applications, it is desirable to store MAVs in small containers before releasing them. Flexible-wing MAVs can be easily reconfigured for storage purposes.

 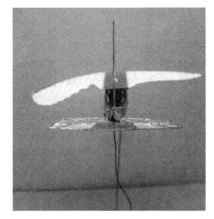

Figure 3.7. The flexible wing allows for wing warping to enhance vehicle agility (courtesy Richard Lind http://128.227.42.147/rick/rick_pro/rick_mav.html).

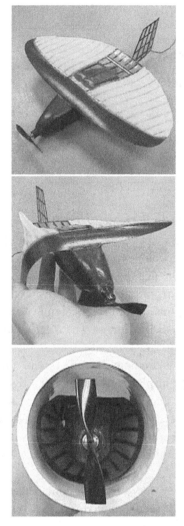

Figure 3.8. A foldable wing to enhance MAV portability and storage (courtesy Peter Ifju).

Figure 3.8 illustrates a 28-cm (11-in.) wingspan foldable-wing MAV that can be stored in a 7.6-cm- (3-in.-) diameter canister. The wing utilizes a curved-shell structural element on the leading edge. This allows the wing to readily collapse downward for storage yet maintain rigidity in the upward direction to react to the aerodynamic loads. The effect is similar to that of a common tape measure, in which the curvature in the metallic tape is used to retain the shape after it has unspooled from the casing yet can be rolled back into the casing to accommodate the small-diameter spool. The curvature ensures that the positive (straight) shape is developed after it is unwound from the case and can actually be cantilevered for some distance. The curvature of the leading edge of the wing acts as the curvature in the tape measure.

Compared with that of rigid-wing aerodynamics, research on flexible-wing aerodynamics is far less extensive. In this chapter, we first present the two flexible-wing

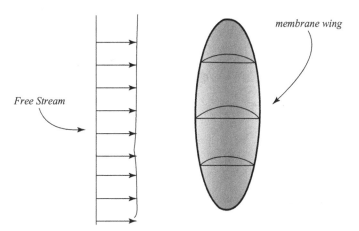

Figure 3.9. Schematic of membrane wing of finite span operating in a free-stream (Shyy and Smith, 1997).

models used; then we discuss the coupled fluid–structural interactions for fixed, flexible wings. Flapping-wing aerodynamics is presented in the next chapter.

3.2 Flexible-Wing Models

In this section, we first review some salient features of membrane-wing dynamics, including the scaling parameters, and one linear and one nonlinear structural model. Then we offer a brief discussion of the computational efforts for the combined fluid and structural dynamics.

3.2.1 *Linear Membrane Model*

As an illustration, consider the membrane wing of Figure 3.9, which is shown operating in a free-stream. A major interest is to probe the coupled dynamics between the fluid flow and the flexible structure. The fluid flow creates pressure and viscous stresses, which cause the membrane to deform. The membrane, in turn, affects the fluid flow structure by means of the shape change, resulting in the so-called moving-boundary problem (Shyy et al., 1996).

The analysis of membrane wings begins with the historical works of Voelz (1950), Thwaites (1961), and Nielsen (1963). These works consider the steady, 2D, irrotational flow over an inextensible membrane with slack. As a consequence of the inextensible assumption and the additional assumptions of small camber and incidence angle, the membrane-wing boundary-value problem is linearized and may be expressed compactly in nondimensional integral equation form as

$$1 - \frac{C_T}{2} \int_0^2 \frac{\frac{d^2(y/\alpha)}{d\zeta^2}}{2\pi(\zeta - x)} d\zeta = \frac{dy/\alpha}{dx}, \tag{3.1}$$

where $y(x)$ defines the membrane profile as a function of the x coordinate, α is the flow incidence angle, C_T is the tension coefficient, and ζ is the arc length along

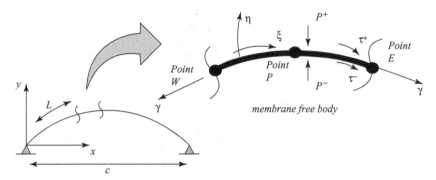

Figure 3.10. Leading-edge-constrained elastic membrane.

the membrane-wing surface. Equation (3.1) was referred to as the "Thwaites sail equation" by Chambers (1966) and simply as the "sail equation" by Greenhalgh et al. (1984) and Newman (1987). This equation, together with a dimensionless geometric parameter ε, completely defines the linearized theory of an inextensible membrane wing in a steady, inviscid flow field. Parameter ε specifies the excess length of an initially flat and taut membrane and is defined as follows:

$$\varepsilon = \frac{L_0 - c}{c}, \tag{3.2}$$

where L_0 is the unstrained length of the membrane and c is the chord length. The meaning of these aforementioned symbols can be better understood from Figure 3.10.

Different analytical and numerical procedures have been applied to the basic equation set in order to determine the membrane shape, aerodynamic properties, and membrane tension in terms of the AoA and excess length. In particular, Thwaites (1961) obtained eigensolutions of the sail equation that are associated with the wing at an ideal angle of incidence. Nielsen (1963) obtained solutions to the same equation by using a Fourier series approach that is valid for wings at angles of incidence other than the ideal angle. Other more recent but similar works are those by Greenhalgh et al. (1984), Sugimoto and Sato (1988), and Vanden-Broeck and Keller (1981).

Various extensions of the linear theory have appeared in the literature over the years. Vanden-Broeck (1982) and Murai and Maruyama (1980) developed nonlinear theories that are valid for large camber and incidence angle. The effect of elasticity was included in the membrane-wing theories of Jackson (1983) and Sneyd (1984), and the effects of membrane porosity were investigated by Murata and Tanaka (1989). In a paper by de Matteis and de Socio (1986), experimentally determined separation points were used to modify the lifting potential-flow problem in an attempt to model flow separation near the trailing edge. A comprehensive review of the work published before 1987 related to membrane-wing aerodynamics was given by Newman (1987).

The agreement between the various potential-flow-based membrane-wing theories and experimental data has been reported by several authors including Greenhalgh et al. (1984), Sugimoto and Sato (1988), and Newman and Low (1984). In general, there has been considerable discrepancy between the measurements made by different authors (Jackson, 1983), which have all been in the turbulent-flow regime at Reynolds numbers between 10^5 and 10^6. As a result of the discrepancies in the reported data, which are primarily due to differences in Reynolds number and experimental procedure, the agreement between the potential-based membrane theories and the data is mixed. In particular, the measured lift is in fair agreement with the predicted value when the excess length ratio is less than 0.01 and the AoA is less than $5°$. However, even for this restricted range of values, the measured tension is significantly less than that predicted by theory. Furthermore, for larger excess lengths and incidence angles the lift and tension are poorly predicted by the theory.

The main reason for the disagreement is the existence of the viscous effect, which significantly affects the force distribution on the wing and therefore the effective shape of the wing. To illustrate the flexible structural dynamics in response to aerodynamic forces, consider equilibrium equations for a 2D elastic membrane subjected to both normal and shear stresses. As discussed in Shyy et al. (1996), the membrane is considered to be massless, and the equilibrium conditions are stated in terms of the instantaneous spatial Cartesian coordinates and the body-fitted curvilinear coordinates. The basic formulation is essentially identical to many previously published works such as those of de Matteis and de Socio (1986) and Sneyd (1984).

Figure 3.10 illustrates an elastic membrane restrained at the leading and trailing edges and subjected to both normal and tangential surface tractions p and τ, respectively. Imposing equilibrium in the normal and tangential directions requires that

$$\frac{d^2 y}{dx^2} \left[1 + \left(\frac{dy}{dx} \right)^2 \right]^{-\frac{3}{2}} = -\frac{\Delta p}{\gamma}, \tag{3.3}$$

$$\frac{d\gamma}{d\xi} = -\tau, \tag{3.4}$$

where γ is the membrane tension. Equation (3.3) is the Young–Laplace equation cast in Cartesian coordinates. The net pressure and shear stress acting on a segment of the membrane are given respectively by

$$\Delta p = p^- - p^+, \tag{3.5}$$

$$\tau = \tau^- - \tau^+, \tag{3.6}$$

where the superscripts indicate the values at the upper and the lower surfaces of the membrane, as shown in the figure. If the membrane material is assumed to be linearly elastic, the nominal membrane tension $\bar{\gamma}$ may be written in terms of the nominal membrane strain $\bar{\delta}$ as

$$\bar{\gamma} = (S^0 + E\bar{\delta})h, \tag{3.7}$$

where S^0 is the membrane prestress, E is the elastic modulus, and h is the membrane thickness. The nominal membrane strain is given by

$$\bar{\delta} = \frac{L - L_0}{L_0},\tag{3.8}$$

where L_0 is the unstrained length of the membrane and L is the length of the membrane after deformation, which may be expressed in terms of the spatial Cartesian coordinates as

$$L = \int_0^c \sqrt{1 + \left(\frac{dy}{dx}\right)^2}\, dx,\tag{3.9}$$

where c is the chord length.

The aeroelastic boundary-value problem can be written in nondimensional form after the following dimensionless variables are introduced:

$$X = \frac{x}{c},\tag{3.10}$$

$$Y = \frac{y}{c},\tag{3.11}$$

$$P = \frac{p}{\rho U_{\mathrm{ref}}^2} = \frac{p}{q_\infty},\tag{3.12}$$

$$\hat{\gamma} = \frac{\gamma}{S^0 h},\tag{3.13}$$

or

$$\hat{\gamma} = \frac{\gamma}{Eh},\tag{3.14}$$

where either Eq. (3.13) or Eq. (3.14) is used to nondimensionalize the membrane tension, depending on whether the tension is dominated by pretension or by elastic strain. The resulting dimensionless equilibrium equation when membrane tension is dominated by elastic strain is

$$\frac{d^2 Y}{dX^2}\left[1 + \left(\frac{dY}{dX}\right)^2\right]^{-\frac{3}{2}} = -\left(\frac{1}{\Pi_1}\right)\frac{\Delta P}{\hat{\gamma}},\tag{3.15}$$

with Π_1 defined as

$$\Pi_1 = \frac{Eh}{q_\infty c}.\tag{3.16}$$

When membrane tension is dominated by pretension, Eq. (3.3) leads to the following dimensionless equation:

$$\frac{d^2 Y}{dX^2}\left[1 + \left(\frac{dY}{dX}\right)^2\right]^{-\frac{3}{2}} = -\left(\frac{1}{\Pi_2}\right)\frac{\Delta P}{\hat{\gamma}},\tag{3.17}$$

with Π_2 defined as

$$\Pi_2 = \frac{S^0 h}{q_\infty c}. \tag{3.18}$$

If the two ends of a 2D membrane are fixed, the boundary conditions in dimensionless form are

$$Y = 0 \ \text{at} \ X = 0, 1 \tag{3.19}$$

Regarding the physical significance of the aeroelastic parameters Π_1 and Π_2, we note that the dimensionless deformation of an initially flat elastic membrane is inversely proportional to Π_1 in the absence of pretension. Alternatively, the dimensionless deformation of a membrane is inversely proportional to Π_2 in the presence of large initial pretension. Consequently, the steady-state, inviscid aeroelastic response of an initially flat membrane wing at a specified AoA is controlled exclusively by Π_1 in the limit of vanishing pretension and exclusively by Π_2 in the limit of vanishing material stiffness.

The preceding scaling analysis is based on a massless structure. If the airfoil mass is considered, then the inertia scaling needs to be considered. Between the elastic and inertia scaling, one can also deduce the structural natural frequency.

3.2.2 *Hyperelastic Membrane Model*

A rubberlike material can be used to cover the rigid skeleton of the MAV design to obtain the wing's flexibility (Figure 3.6). The large deformations observed for this kind of material in the Reynolds number range of operation indicate that the linear elasticity assumption may not be valid.

To address this issue, a hyperelastic model to describe the 3D membrane material behavior is used (Lian et al., 2003a). The stress-strain curve of a hyperelastic material is nonlinear, but follows the same path in loading and unloading (below the plastic limit, which is significantly higher than in metals). Compared with the previously discussed 2D linear model, a 3D membrane model introduces several complicated factors. First, for 3D membranes, the tension is defined as a biaxial tension along the lines of principal stress (Jackson and Christie, 1987). Second, the geometric and material properties may vary along the spanwise direction and need to be described in detail. A third factor is membrane compression, which leads to wrinkles when one of the principal tensions vanishes. In addition, it is desirable to account for the membrane mass when solving for the dynamic equations of the membrane movement.

A finite-element analysis of the static equilibrium of an inflated membrane undergoing large deformations is presented by Oden and Sato (1967). A review of the earlier work on the dynamic analysis of membranes can be found in Jenkins and Leonard (1991). An update of the state-of-the-art models in membrane dynamics is presented by Jenkins (1996). Verron et al. (2001) studied, both numerically and experimentally, the dynamic inflation of a rubberlike membrane. Ding et al. (2003)

numerically studied partially wrinkled membranes. In a recent effort, Stanford et al. (2006) proposed a linear model for 3D membranes used in MAV design. Their experimental measurements showed that the maximum strain value is quite small; therefore a linear approximation of the stress-strain curve is constructed, centered about the membrane wing's prestrain value. The linear constitutive equation used for membrane modeling is Poisson's equation:

$$\frac{\partial^2 W}{\partial x^2} + \frac{\partial^2 W}{\partial y^2} = -\frac{p(x, y)}{S}, \tag{3.20}$$

where W is the out-of-plane membrane displacement, p is the applied pressure (wind loading, in this case), and S is the membrane tension per unit length. The aerodynamic loads are computed on a rigid wing and fed into the structural model, assuming that the change in shape of the membrane wing did not overtly redistribute the pressure field. Good agreement is obtained between the experimental data and computations.

A 3D membrane model was developed by Lian et al. (2003a). The model gives good results for membrane dynamics with large deformations but has a limited capability to handle the wrinkle phenomenon that occurs when the membrane is compressed. The membrane material considered obeys the hyperelastic Mooney–Rivlin model (Mooney, 1940). A brief review of their membrane model is given next.

The Mooney–Rivlin model is one of the most frequently used hyperelastic models because of its mathematical simplicity and relatively good accuracy for reasonably large strains (less than 150%) (Mooney, 1940). For an initially isotropic membrane, a strain-energy function W can be defined as (Green and Adkins, 1960):

$$W = W(I_1, \ I_2, \ I_3), \tag{3.21}$$

where I_1, I_2, and I_3 are the first, second, and third invariants of the Green deformation tensor $\mathbf{C}$, respectively. For an incompressible material, when $I_3 = 1$, the strain energy is a function of I_1 and I_2 only, and a linear expression can be written for the membrane strain energy:

$$W = c_1 \left(I_1 - 3\right) + c_2 \left(I_2 - 3\right), \tag{3.22}$$

where c_1 and c_2 are two material constants. A material that obeys Eq. (3.22) is known as a Mooney–Rivlin material. For an initially isotropic membrane, the general stress-strain relation is written as

$$\mathbf{S} = -p\mathbf{C}^{-1} + 2\frac{\partial W}{\partial \mathbf{C}}, \tag{3.23}$$

where $\mathbf{S}$ is the second Piola–Kirchhoff stress tensor and p is the hydrostatic pressure. If the membrane is incompressible, the stress-strain relation can be simplified further:

$$\mathbf{S} = -p\mathbf{C}^{-1} + 2\left[(c_1 + c_2 I_1) \cdot \mathbf{I} - c_2 \mathbf{C}\right], \tag{3.24}$$

where $\mathbf{I}$ is the 3×3 identity matrix. The membrane stress field is essentially assumed to be 2D, and therefore the stress normal to the deformed membrane surface is negligible with respect to the stress in the tangent plane (Oden and Sato, 1967). Under this assumption, the deformation matrix $\mathbf{C}(t)$ and the stress matrix $\mathbf{S}(t)$ can be written as

$$\mathbf{C}(t) = \begin{bmatrix} C_{11}(t) & C_{12}(t) & 0 \\ C_{12}(t) & C_{22}(t) & 0 \\ 0 & 0 & C_{33}(t) \end{bmatrix}; \quad \mathbf{S}(t) = \begin{bmatrix} S_{11}(t) & S_{12}(t) & 0 \\ S_{12}(t) & S_{22}(t) & 0 \\ 0 & 0 & 0 \end{bmatrix}. \quad (3.25)$$

The hydrostatic pressure is determined by the condition that $S_{33} = 0$, and the formula is

$$p = 2\left(c_1 + c_2 I_1 - c_2\lambda_3^2\right)\lambda_3^2, \quad (3.26)$$

where λ_3 is defined by

$$\lambda_3 = \sqrt{C_{33}(t)} = \frac{h(t)}{h_0}, \quad (3.27)$$

in which $h(t)$ and h_0 are the membrane thickness in the deformed and nondeformed configurations, respectively.

More details about the model, validation, and numerical implementation can be found in Lian and Shyy (2005), Lian et al. (2003b), and Verron et al. (2001).

3.2.3 *Combined Fluid–Structural Dynamics Computation*

Computational Fluid Dynamics (CFD) To analyze the flow field under consideration, the Navier–Stokes equations represent the fluid dynamics aspect of flexible-wing dynamics. The solution techniques for such a problem involving moving coordinates are given in Shyy et al. (1996). For moving-boundary problems in which a solid boundary (e.g., wing) moves inside a computational domain based on known kinematics (e.g., a flapping wing, which will be discussed in Chapter 4) or as a response of the structure to the flow around it (e.g., a fixed, flexible wing), the grid needs to be adjusted dynamically during computation. To facilitate this, as presented by Shyy et al. (1996), either a moving-grid technique or a fixed grid can be used. A fixed-grid technique (Dong et al., 2006; Singh et al., 2006; Ye et al., 2001) is attractive because no regridding is needed. Furthermore, an adaptive, local grid-refinement technique can be adopted (Singh and Shyy, 2006). For flow computations involving transition but no merger or breakup of the multiple objects, which is the case for MAV aerodynamics, the disparity in length scales coupled with moving objects can be better treated with a moving-grid technique.

In the moving-grid approach, the process of generating a grid can be a complicated task by itself so an automatic and fast algorithm to upgrade the grid frequently is essential. It is desirable to have an automatic remeshing algorithm to ensure that the dynamically moving grid retains the quality of the initial grid and avoids problems such as crossover of the grid lines, crossed cell faces, or negative volumes at block interfaces in the case of multiblock grids. Several approaches have been developed to treat grid redistributions for moving-grid computations (Lian et al., 2003b).

Computational Structural Dynamics (CSD) In additition to the membrane models just reviewed, as further discussed in Chapter 4, insect/bird wings are made of a combinations of veins/bones and membranes/feathers. Using these basic constructions as a starting point, the structural model needs to account for anisotropy and large deformations and to be composed of multistructural components. To achieve the modeling flexibility desirable and compatible with the CFD fidelity, the combination of localized stiff members and thin unreinforced skin can be modeled as a combination of beam, wires, plates, and membranes, along with their distributed inertia. Each of these elements (and therefore their corresponding governing equations) will be capable of orthotropic material properties and large motions. The membrane and wiring elements should also be able to handle stress stiffening.

Coupling of CFD and CSD To model the flexible-wing performance, the fluid dynamics and structural dynamics models need to be computed with coordination. Partitioned analyses have been very popular in the area of computational fluid–structure interactions/computational aeroelasticity. A main motivating factor in adopting this approach is that one can develop and use state-of-the-art fluid and structure solvers and recombine them with minor modifications to allow for the coupling of the individual solvers. The accuracy and stability of the resulting coupled scheme will depend on the selection of the appropriate interface strategy, which depends on the type of application. The key requirements for any dynamic coupling scheme are (i) kinematic continuity of the fluid–structure boundary, which leads to the mass conservation of the wetted surface; (ii) dynamic continuity of the fluid–structure boundary, which accounts for the equilibrium of tractions on either side of the boundary. This leads to the conservation of linear momentum of the wetted surface. Energy conservation at the fluid–structure interface requires that both of the preceding continuity conditions be satisfied simultaneously.

The subject of fluid and structural interactions is vast. Recent reviews by Friedmann (1999) and Livne (2003) offer substantial information and references of interest to us.

3.3   Coupled Elastic Structures and Aerodynamics

3.3.1 *Flexible Airfoils*

Shyy et al. (1997) first reported the relative performance between a low Reynolds number membrane and a rigid airfoil in terms of the lift-to-drag ratio. Both airfoils are of the same nominal camber at 6% in a fluctuating free stream. To mimic the effect of wind gust, the fluctuations in the free stream can be modulated up to 25% or more, and the varying free stream has a sinusoidal modulation frequency set at 1.7 Hz. In addition, a hybrid airfoil was also investigated, which is a combination of rigid and membrane airfoils. The linear membrane model was used to account for the airfoil's flexibility. This airfoil was built with a curved-wire screen beneath the membrane. By adjusting the hybrid airfoil, one achieves a camber of approximately

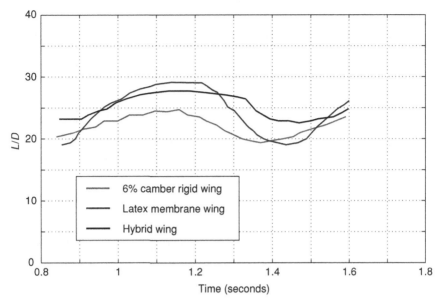

Figure 3.11. Experimental L/D results for rigid, flexible, and hybrid wings at $Re = 7.5 \times 10^4$ and an AoA of $7°$. The latex membrane wing exhibits about 6% camber at 35.4 fps. The hybrid wing has a curved-wire screen camber stop. From Shyy et al. (1997).

6% for this configuration as well. The hybrid airfoil can hence achieve a camber greater than 6%, but not less because the wire construction prevents a decrease in the camber. The size of the wing chords and the average wind-tunnel speed give a Reynolds number of 7.5×10^4.

The experiments for all three airfoils were conducted at an AoA of $7°$, and the results are shown in Figure 3.11. Detailed numerical simulations based on the Navier–Stokes equations and two-equation turbulence closure, along with a moving-grid technique to track the shape variations of the membrane and hybrid airfoils, are conducted for various cases. At modest AoAs, the airflow over the rigid airfoil is attached to the surface at all times and the lift-to-drag ratio follows the free-stream fluctuations. When the AoA is increased ($\alpha = 7°$) the flow separation can become very substantial, causing a modification to the effective shape of the rigid airfoil. As the AoA is increased, the lift coefficient C_L tends to increase as well, but the lift-to-drag ratio decreases because of flow separation. The separation at higher AoAs makes the airfoil less sensitive to an unsteady free stream. For both AoAs, the lift coefficients between rigid and flexible airfoils are comparable, but the lift-to-drag ratio is higher for the flexible airfoil. For the membrane airfoil, at $\alpha = 7°$ the flow separation is confined to the leading edge, resulting in the better aerodynamic performance. There are, however, some negative effects with a flexible membrane. When the free-stream velocity reaches its lower value during a fluctuating cycle, the camber of the flexible membrane tends to collapse and a massive separation over the whole surface occurs. This phenomenon is due to the smaller pressure differences between the upper and

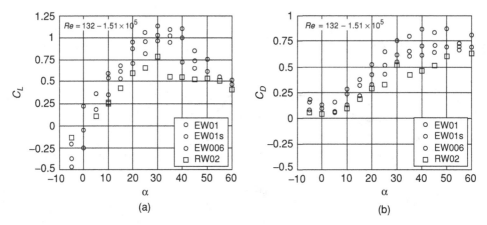

Figure 3.12. Lift and drag coefficients over rectangular wings with different flexibilities: RW02, a rigid steel-plate wing; EW006 and EQ01, thin and thick latex membranes; EW01s, a latex membrane with 6% slack (Galvao et al., 2006).

the lower surface of the membrane and hence a degraded performance is obvious. The hybrid airfoil, which has a curved-wire screen stop to prevent the camber from becoming too low, shows interesting results. For a lower AoA, the aerodynamic characteristics are about the same as those for the flexible airfoil, with an attached airflow. When the AoA is increased, the characteristics are considerably better for the hybrid airfoil compared with the flexible profile. The separation zone is smaller compared with that of the rigid airfoil, and the sensitivity to fluctuations in the free stream is reduced when compared with that of the flexible configuration.

3.3.2 *Membrane-Wing Aerodynamics*

Galvao et al. (2006) measured the lift, drag, and deflection of a compliant rectangular membrane wing at a Reynolds number range of 7×10^4 to 2×10^5 over a range of AoAs (−5°–60°). The wing is composed of a compliant latex membrane held between two stainless steel posts located at the leading and the trailing edges. Four wing models are tested: a thin noncompliant wing composed of a steel shim stock (denoted RW02), two compliant membrane wings with latex rubber sheets of thicknesses of 0.25 and 0.15 mm (denoted EW01 and EW006, respectively), and a latex membrane wing (0.25 mm thick) in which the membrane is given 6% slack (denoted EW01s).

Figure 3.12(a) depicts the lift coefficient for the test. The compliant wings have a greater lift slope than the rigid wing. And the thinner compliant wing has a greater lift slope than the thick compliant wing. Wing-deflection measurements show that this is due to the increased camber for the compliant wing, which is consistent with the numerical results (Lian and Shyy, 2003). The thinner compliant wing stretches to a greater degree compared with the thicker membrane wing and therefore has a larger camber at same AoA, resulting in a larger lift coefficient. Whereas Figure 3.5 shows that membrane wings have a lift slope similar to that of the rigid wing, Figure 3.12(a)

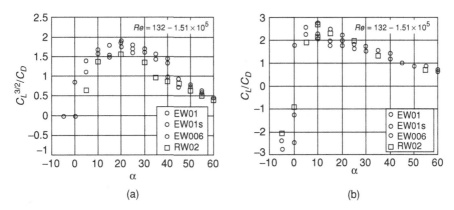

Figure 3.13. Measured power efficiency and lift-to-drag ratio over four wings with different flexibilities: RW02, a rigid steel-plate wing; EW006 and EQ01, thin and thick latex membranes; EW01s, a latex membrane with 6% slack (Galvao et al., 2006).

indicates that flexible wings have a greater lift slope than rigid wings. The seemingly contradicting conclusion is due to different experimental setups.

Figure 3.5 is based on the measurement of an MAV with a free trailing edge, which can be tilted up under forces (Waszak et al., 2001). As pointed out by Lian and Shyy (2005), this trailing-edge deflection reduces the effective AoA. The interplay among the camber, effective AoA, and the lift can be rather complicated. Before stall, the flexible structure in the experiment of Waszak et al. exhibits a smaller effective AoA; because the trailing edge is not fixed, the effective camber is reduced accordingly. On the other hand, in the experiment of Galvao et al. (2006), the trailing edge is fixed, resulting in a fixed AoA but higher camber.

Galvao et al. reported that a compliant wing can delay stall by $2°$ to $8°$ of AoA, which is qualitatively consistent with the observation of Waszak et al. (2001). After stall, the lift coefficients for the compliant wings decrease in a more attenuated manner compared with that of the rigid wing. Close to the stall, the camber of the wing is observed to decrease. The decambering acts decrease the severity of the separation, thus delaying the sharp drop in lift force. This behavior enables the wing to sustain high lift at high AoAs. Furthermore, the compliant wings generate more lift at AoAs from $5°$ to $55°$.

The compliant wings are found to yield more drag (Figure 3.12) for two possible reasons. First, the enlarged camber increases the form drag. Second, the high-frequency fluctuation and vibration heighten the drag. This becomes more noticeable when the trailing edge is not fixed, possibly leading to flutter. As we explained before, during slow forward flight, bats can flex their wings only slightly to avoid flutter. The compliant wings also demonstrate their superiority in terms of power efficiency ($C_L^{3/2}/C_D$) over a wide range of AoAs [Figure 3.13(a)] (Galvao et al., 2006). This becomes more evident at higher AoAs. However, in terms of flight range efficiency (C_L/C_D), compliant wings have a performance comparable with that of the rigid wing [Figure 3.13(b)].

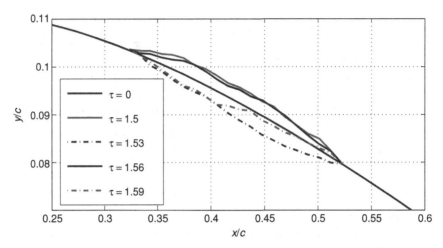

Figure 3.14. Membrane airfoil shapes in a steady free stream at different time instants. The vibration changes the effective wing camber, where τ is the nondimensional time, defined as tc/U (Lian and Shyy, 2006).

Lian and Shyy (2006) numerically investigated flexible-airfoil aerodynamics. In their test, the upper surface of the airfoil is covered with a membrane that extends from 33% to 52% of the chord. No pretension is applied to the membrane. The membrane has a uniform thickness of 0.2 mm with a density of 1200 kg/m^3. The two parameters governing the membrane property, as shown in Eq. (3.22), take the values of $c_1 = 5.0 \times 10^5$ Pa and $c_2 = 0.785c_1$. The reference scales of their computations are based on the free-stream velocity of 0.3 m/s, a density of 1000 kg/m^3, and an airfoil chord length of 20 cm. With these parameters, the time step for the CFD solver here is set to 2×10^{-3} s and the time step of the structural solver is 1×10^{-5} s. The structural solver is very fast, and the majority of the CPU time is for the CFD solver. The use of iteration between the CFD solver and structural solver during each time step allows for synchronization of the fluid and structure coupling. By doing this the errors introduced by a lagged fluid–structure coupling approach are regulated.

A computational test is performed at $\alpha = 4°$ and $Re = 6 \times 10^4$. It is observed that, when flow passes the flexible surface, the surface experiences self-excited oscillation and the airfoil displays varied shapes over time (Figure 3.14). Analysis shows that the transverse velocity magnitude can reach as much as 10% of the free-stream speed. During the vibration, energy is transferred from the wall to the flow and the separated flow is energized. Compared with the corresponding rigid-airfoil simulation, the surface vibration causes both the separation and transition positions to exhibit a standard variation of 6% of the chord length.

In Figure 3.15 the time history of the lift coefficient is presented. Even though the time-averaged lift coefficient (0.60) of a flexible wing is comparable with that of the corresponding rigid wing, the lift coefficient displays a time-dependent variation with maximum magnitude as much as 10% of its mean. The drag coefficient shows a similar pattern but the time-averaged value closely matches that of the rigid wing.

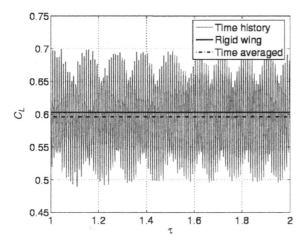

Figure 3.15. Time history of lift coefficient for membrane wing, showing both a high-
and a low-frequency oscillation (Lian and Shyy, 2006).

These observations are consistent with our previous efforts in 3D MAV wing simulations, without transitional flow models (Lian and Shyy, 2005). Furthermore, the
experimental evidence also supports the fact that, until the stall condition is reached,
the membrane and the rigid wings exhibit comparable aerodynamic performances.
The flexible wing, on the other hand, can delay the stall margin substantially (Galvao
et al., 2006; Waszak et al., 2001). By use of discrete Fourier transformation analysis,
the primary frequency of this flexible *airfoil* is found to be 167 Hz (Figure 3.16).
Given the airfoil chord (0.2 m) and free-stream speed (0.3 m/s), this high-frequency
vibration is unlikely to affect the vehicle stability. Figure 3.15 indicates that a low-
frequency cycle exists in the high-frequency behavior in the lift coefficient history.

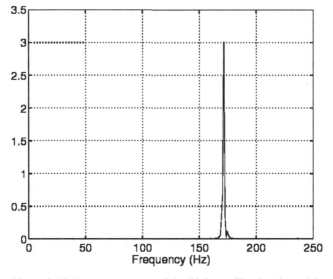

Figure 3.16. Power spectrum of the lift force. The dominated frequency is 167 Hz (Lian
and Shyy, 2006).

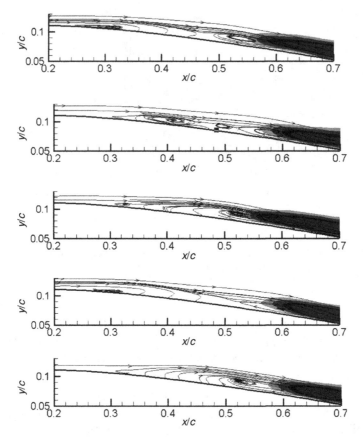

Figure 3.17. Flow structure over the membrane wing and the associated vortex shedding at $\alpha = 4°$ and $Re=6 \times 10^4$. From top to bottom, the time instant $\tau =1.5$, 1.506, 1.512, 1.515, 1.521 (Lian and Shyy, 2006).

This cycle, with a frequency of about 14 Hz, seems to be associated with vortex shedding (Figure 3.17). In a different simulation with laminar flow over a 6-in. (15.24-cm) membrane *wing* (i.e., the entire wing surface is flexible), Lian and Shyy observed a self-excited structural vibration with a frequency of around 120 Hz (Lian and Shyy, 2005); the experimental measurement of similar wings records a primary frequency of around 140 Hz (Waszak et al., 2001).

Lian et al. (2003b) compared the aerodynamics between membrane and rigid wings for MAV applications. The flexible wing exhibits a slightly smaller lift coefficient than the rigid one at $\alpha = 6°$. The difference in C_L/C_D is also small. At a higher AoA of 15°, the membrane wing generates a lift coefficient about 2% smaller than that of the rigid wing; however, its C_L/C_D is slightly larger than that of the rigid wing. This observation is consistent with the findings of Shyy et al. (1997).

The membrane wing changes its shape under external forces. This shape change has two effects. On the one hand, it decreases the lift force by reducing the effective AoA of the membrane wing; on the other hand, it increases the lift force by increasing the camber. Both numerical findings of Lian and Shyy (2003) and experimental

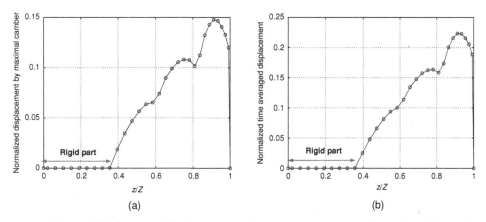

Figure 3.18. Averaged displacement of the membrane-wing trailing edge: (a) $\alpha = 6°$, (b) $\alpha = 15°$. Adopted from Lian and Shyy (2003).

observations of Waszak et al. (2001) have shown that membrane and rigid wings exhibit comparable aerodynamic performances before the stall limit.

Figure 3.18 shows the time-averaged vertical displacement of the trailing edge. The displacement is normalized by the maximal camber of the wing. Because of membrane deformation, the effective AoA of the membrane wing is less than that of the rigid wing. The spanwise AoAs between rigid and membrane wings under the same flow condition and with identical initial geometric configurations are shown in Figure 3.19. In Figure 3.19(a), the rigid wing has an incidence of 6° at the root and monotonically increases to 9.5° at the tip; the membrane wing shares the same AoAs with the rigid wing at 36% of the inner wing; however, the effective AoA toward the tip is less than that of the rigid wing. At the tip, the AoA of the membrane wing lowers by about 0.8°. Figure 3.19(b) compares the AoA at $\alpha = 15°$; it shows that the effective AoA of the membrane wing is more than 1° less than that of the rigid wing at the tip. The reduced effective AoA causes the decrease in the lift force.

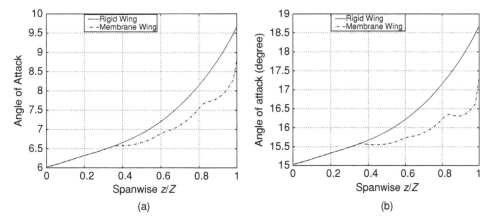

Figure 3.19. Time-averaged spanwise AoA for membrane wing: (a) $\alpha = 6°$, (b) $\alpha = 15°$. Adopted from Lian and Shyy (2003).

3.4   Concluding Remarks

In this chapter, we discussed the interplay between structural flexibility and aerodynamics. Representative modeling concepts were described, and the computational strategies for the coupled fluid–structural dynamics were highlighted:

1. Flexible wings are found to be beneficial for both natural and manmade flyers. Birds can flex their wings during upstroke to minimize the drag and can still maintain a smooth surface by slipping the features together. Bats, whose wings consist of membrane and arm bones, can flex the wings only a bit to avoid structure failure or flutter; however, they can enlarge the wing camber during the downstroke; insects can bend the wing chordwise to generate camber while preventing bending in the spanwise direction.

2. Fixed, flexible wings can facilitate steadier and better controlled flight. In a gusty environment, a flexible wing can provide a more consistent lift-to-drag ratio than a rigid wing by adaptively adjusting the camber in accordance with the instantaneous flow field.

3. By responding to the aerodynamic loading variations, a membrane wing can also adaptively conduct passive camber control to delay stall.

4. A membrane wing is found to exhibit flutter, whose frequency is about an order of magnitude higher than that of the vortex-shedding frequency. The flutter exists even under a steady-state free-stream condition. Such intrinsic vibrations result from coupled aerodynamics and structural dynamics.

CHAPTER FOUR

Flapping-Wing Aerodynamics

Flying animals flap wings to create lift and thrust as well as to perform remarkable maneuvers with rapid accelerations and decelerations. Insects, bats, and birds provide illuminating examples of utilizing unsteady aerodynamics to design future MAVs.

Pioneering work on flapping-wing aerodynamics was done by Lighthill (1969) and Weis-Fogh (1973). Recent works, both in experiments and simulations, were documented by Katz (1979), Ellington (1984a), DeLaurier (1993), Smith (1996), Vest and Katz (1996), Liu and Kawachi (1998), Dickinson et al. (1999), Jones and Platzer (1999, 2003), Wang (2000), and Chasman and Chakravarthy (2001). A review of the characteristics of both flapping wings and fixed wings was given by Shyy et al. (1999a). The spectrum of animal flight with flapping wing was presented by Templin (2000). Ho et al. (2003) further reviewed the recent effort in developing flapping-wing-based MAVs. Computational and experimental studies regarding rotating-wing MAVs were made by Bohorquez et al. (2003).

Aerodynamic phenomena associated with biological flight prominently features unsteady motions, characterized by large-scale vortex structures, complex flapping kinematics, and flexible-wing structures. Furthermore, knowledge gained from studying biological flight shows that the steady-state aerodynamic theory can be seriously challenged to explain the lift needed for biological flyers (Brodsky, 1994; Ellington, 1984a; Ellington et al., 1996).

The quasi-steady theory is constructed based on the instantaneous velocity, wing geometry, and AoA when the steady-state aerodynamic model is used. By neglecting the wing motion and flow history, the quasi-steady approach greatly simplifies the time-dependent problem by converting it to a sequence of independent, steady-state problem: It has been frequently used in interpreting biological flight characteristics (Azuma, 1983; Lighthill, 1973; Maxworthy, 1979; Norberg, 1990; Pennycuick, 1989; Spedding, 1992; Weis-Fogh, 1972). For example, it has been used to the estimate mechanical power requirements of hummingbirds (Chai and Dudley, 1996) and bumblebees (Dudley and Ellington, 1990b). Based on the theoretical analyses (Ellington, 1995) and experimental measurements of tethered insects (Cloupeau, 1979; Wilkin and Williams, 1993), it has been found that the quasi-steady model is insufficient to predict the lift needed to support insect body weight. On the other hand, while investigating a dynamically scaled, rigid-winged, flapping insect mounted in mineral oil (*Robofly*), several authors (Sane and Dickinson, 2001; Wang et al., 2004) suggested that the quasi-steady 2D blade-element models can yield satisfactory agreement with the experimental measurement of aerodynamic forces. We will visit this issue in later sections. In any event, it is clear that the advancements in laser diagnostics and other experimental techniques, robotics and control, and CFD and CSD have fostered a fruitful collaboration in flapping-wing research (Combes and Daniel, 2003; Dickinson et al., 1999; Ellington et al., 1996; Liu and Kawachi, 1998; Sunada et al., 2001; Wang, 2000).

In this chapter we present the various issues related to the aerodynamics of flapping flight. We first discuss the scaling of flapping-wing flight in terms of reduced frequency, Reynolds number, Strouhal number, and advance ratio. We next review nonstationary airfoil aerodynamics including dynamic stall, a plunging and pitching airfoil, and thrust generation. Both analytical and computational models are highlighted. Then we discuss the main lift-enhancement mechanisms associated with flapping wings, including leading-edge vortex (LEV), fast pitch-up, wake capture, and clap-and-fling mechanisms. The interplay among sizing (Reynolds number), kinematics, and reduced frequency on aerodynamic characteristics is addressed.

4.1   Scaling, Kinematics, and Governing Equations

Aerodynamics of insect and bird flapping flight can be modeled within the framework of unsteady, Navier–Stokes equations. Nonlinear physics with multiple variables (velocity, pressure) and time-varying geometries are among the aspects of primary interest. The treatment of flapping kinematics is therefore central to a comprehensive understanding of animal flight.

Scaling laws can help identify the physical flow regimes as well as offer guidelines to establish suitable models for predicting the aerodynamics of biological flight. Three main dimensionless parameters in flapping-flight scaling are (i) the Reynolds number, which represents a ratio of inertial and viscous forces, (ii) the Strouhal number for forward flight, which describes the relative influence of forward versus flapping speeds, and (iii) the reduced frequency, which describes the rotational versus translational speeds during flapping movement. Together with geometric and kinematic similarities, the Reynolds number, the Strouhal number, and the reduced frequency are sufficient to define the aerodynamic similarity for a *rigid* wing.

4.1.1 *Flapping Motion*

The kinematics of insect flapping flight, depicted in Figure 4.1, describes wing and body movement. The body kinematics can be represented by the body angle χ (inclination of the body), which is relative to the horizontal plane, and the stroke-plane angle β (indicated by the solid lines), which refers to a plane including the wing base and the wingtips of the maximum and the minimum sweep positions. The body angle and the stroke-plane angle vary in accordance with the flight speed. The

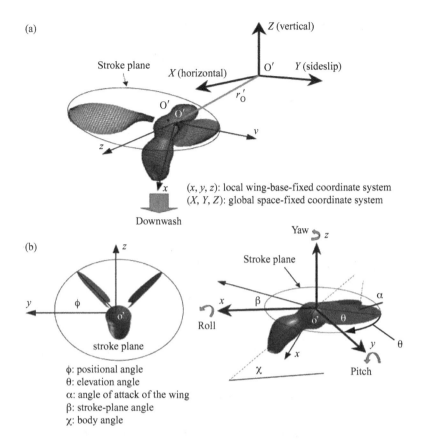

Figure 4.1. Schematic diagram of coordinate systems and wing kinematics: (a) the local wing-base-fixed and the global space-fixed coordinate systems. The local wing-base-fixed coordinate system (x, y, z) is fixed on the center of the stroke plane (origin O' at the wing base) with the x direction normal to the stroke plane, the y direction vertical to the body axis, and the z direction parallel to the stroke plane; (b) definition of the positional angle ϕ, the feathering angle (AoA of wing) α, elevation angle θ of the flapping wing, body angle χ, and stroke-plane angle β.

wing-beat kinematics can be described by three positional angles within the stroke plane: (i) flapping about the x axis in the wing-fixed coordinate system described by the positional angle ϕ, (ii) rotation of the wing about the z axis described by the elevation angle θ, and (iii) rotation (feathering) of the wing about the y axis described by the AoA α.

The AoA α is used to describe the orientation of a chordwise strip of a beating wing relative to the stroke plane, which may change significantly in the spanwise direction because of the wing torsion often observed in insect flapping flight.

For a general 3D case, definitions of the positional angle, the elevation angle, and the AoA, all in radians, are

$$\phi(t) = \sum_{n=0}^{3} [\phi_{cn} \cos(2n\pi ft) + \phi_{sn} \cos(2n\pi ft)], \quad n = \text{integer}, \qquad (4.1)$$

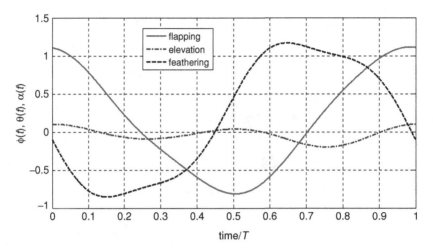

Figure 4.2. Positional-, elevation-, and feathering-angle variations for one period for a hovering hawkmoth.

$$\theta(t) = \sum_{n=0}^{3} [\theta_{cn} \cos(2n\pi ft) + \theta_{sn} \cos(2n\pi ft)], \qquad (4.2)$$

$$\alpha(t) = \sum_{n=0}^{3} [\alpha_{cn} \cos(2n\pi ft) + \alpha_{sn} \cos(2n\pi ft)]. \qquad (4.3)$$

Note that t is time and f is flapping frequency. The Fourier coefficients ϕ_{cn}, ϕ_{sn}, θ_{cn}, θ_{sn}, α_{cn}, and α_{sn} are determined from the empirical kinematic data. Based on the Fourier coefficients gathered by analysis of the kinematics of a hovering hawkmoth (Willmott and Ellington, 1997b), the positional-, elevation-, and feathering-angle variation for one period are plotted in Figure 4.2.

Even though 3D effects are important for predicting low Reynolds number flapping-wing aerodynamics, 2D experiments and computations do provide important insight into the unsteady fluid physics related to flapping wings. Two hovering modes are discussed in this chapter, one is called the "water-treading" mode (Freymuth, 1990) and the other is called the "normal-hovering" mode (Wang et al., 2004). The plunging and pitching of the airfoil are described by symmetric, periodic functions:

$$h(t) = h_a \sin(2\pi ft + \varphi), \qquad (4.4)$$

$$\alpha(t) = \alpha_0 + \alpha_a \sin(2\pi ft), \qquad (4.5)$$

where h_a is the plunging amplitude, f is the plunging frequency, α_0 is the initial rotational angle, α_a is the pitching amplitude, and φ is the phase difference between plunging and pitching motion. The schematics of the two hovering modes are presented in Figure 4.3.

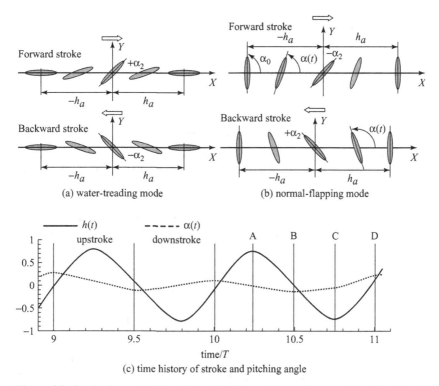

(a) water-treading mode   (b) normal-flapping mode

(c) time history of stroke and pitching angle

Figure 4.3. Basic characteristics of the airfoil hovering modes considered in this study: (a) schematic of water-treading mode; (b) schematic of the normal mode; (c) time histories of airfoil stroke (solid curve, $h(t)$) and pitch angle (dashed curve, $\alpha(t)$) used for both modes.

From the phase relationship between the wing's translation and rotation, Dickinson et al. (1999) further categorized the wing motion into advanced, symmetric, and delayed modes. The advanced mode (Figure 4.4) is the pattern in which the wing rotates before it reverses direction at the end of each stroke. The symmetrical mode (Figure 4.4) is the pattern in which the wing rotation synchronizes with its flapping motion: Its AoA is 90° at the end of each stroke. The delayed mode (Figure 4.4) is the pattern in which the wing rotates after it reverses direction at the end of each stroke.

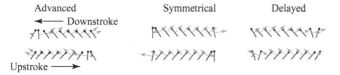

Figure 4.4. Schematics of the three wing-rotation patterns: advanced, symmetrical, and delayed rotation. As shown in Dickinson et al. (1999), the timing of the wing rotation has an important role in lift generation and consequently in maneuvering. Blue designates the airfoil location, with the circle corresponding to the leading edge, and red indicates the resultant aerodynamic force.

4.1.2 *Reynolds Number*

Given a reference length L_{ref} and a reference velocity U_{ref}, one normally defines the Reynolds number Re as

$$Re = \frac{U_{\text{ref}} L_{\text{ref}}}{\nu}, \tag{4.6}$$

where ν is the kinematic viscosity. In biological flight, with consideration of the fact that flapping wings produce the lift and thrust, the mean chord length c_m is used as the reference length L_{ref}, whereas the body length is typically used in swimming animals. The reference velocity U_{ref} is also defined differently in hovering and forward flight:

1. In *hovering*, the mean wingtip velocity may be used as the reference velocity, also written as $U_{\text{ref}} = \omega R$, where R is the wing length (half wing span) and ω is the mean angular velocity of the wing ($\omega = 2\Phi f$, where Φ is the wing-beat amplitude, measured in radians, and f is the flapping frequency). Therefore the Reynolds number for a 3D flapping wing Re_{f3} in hovering flights can be cast as

$$Re_{f3} = \frac{U_{\text{ref}} L_{\text{ref}}}{\nu} = \frac{2\Phi f R c_m}{\nu} = \frac{\Phi f R^2}{\nu} \left(\frac{4}{\text{AR}} \right), \tag{4.7}$$

where the aspect ratio AR as described in Chapter 1 is introduced in the form $\text{AR} = (2R)^2/S$, with the wing area being the product of the wing span $(2R)$ and the mean chord (c_m). Note that the Reynolds number here is proportional to the wing-beat amplitude Φ, the flapping frequency f, a square of the wing length R^2, but inversely proportional to the AR of the wing. In insects flights, the wing-beat amplitude and the AR of the wing do not vary significantly, but the flapping frequency increases as the insect size is reduced, which, in general, results in Re ranging from $O(10^1)$ to $O(10^4)$ (see Figure 1.15). In addition, given a geometrically similar wing model that undergoes flapping hovering with the same wing-beat amplitude, the product of fR^2 can preserve the same Reynolds number. This implies that a scaled-up but low-flapping-frequency wing model can be built mechanically to mimic insect flapping flight based on aerodynamic similarity. In fact, most recent robotic-model-based studies (Dickinson et al., 1999; Ellington et al., 1996) are established on such a basis provided that the second parameter, the Strouhal number, can be satisfied simultaneously.

   The definition of the Reynolds number can also be defined by use of an alternative reference length and/or the reference velocity. For example, with the wing length R as the reference length and the wing velocity $U_{\text{ref}} = \omega r_2 R$, where r_2 is the radius of the second moment of wing area [approximately 0.52 for hawkmoth, *Manduca sexta* (Liu et al., 1998; Van den Berg and Ellington, 1997)], the Reynolds number Re_{f3} is proportional with $\Phi f R^2/\nu$ and is not dependent on the AR of the wing. Note that the reference velocity here is almost half of that at the wingtip.

For a 2D flapping wing (e.g., normal and water-treading hovering modes, discussed later), the Reynolds number Re_{f2} is defined by the maximum plunging velocity:

$$Re_{f2} = \frac{U_{ref}L_{ref}}{\nu} = \frac{2\pi f h_a c}{\nu},$$ (4.8)

where f is the flapping frequency, h_a is the plunging amplitude, and c is the airfoil chord length.

2. In forward flight, for both 2D and 3D applications, the forward velocity U is often used as the reference velocity U_{ref}. Compared with the hovering-flight Reynolds number, which is proportional to R^2, the forward-flight Reynolds number is proportional to R.

4.1.3 Strouhal Number and Reduced Frequency

In flapping wing studies, the Strouhal number (St) is well known for characterizing the vortex dynamics and shedding behavior. In some St ranges, the flapping airfoil produces thrust, and the vortices in the wake are termed *reverse* von Karman vortices. In general, for flapping flight, the dimensionless parameter St describes the dynamic similarity between unsteady systems, and is normally defined as

$$St = \frac{f L_{ref}}{U_{ref}} = \frac{2 f h_a}{U},$$ (4.9)

where f is the stroke (flapping) frequency in flapping flight, h_a is the stroke (flapping) amplitude, and U is the forward velocity. This definition describes a ratio between the oscillating (flapping) speed $(f h_a)$ and the forward speed (U), which offers a measure of propulsive efficiency in flying and swimming animals. In the study of natural flyers and swimmers in cruising condition it is found that the Strouhal number, as defined by Eq. (4.9), is often within a narrow region of $0.2 < St < 0.4$ (Taylor et al., 2003; Triantafyllou et al., 2000).

Another dimensionless parameter that characterizes the unsteady aerodynamics of pitching and plunging airfoils is the reduced frequency, defined as

$$k = \frac{2\pi f L_{ref}}{2U_{ref}} = \frac{\pi f c_m}{U}.$$ (4.10)

In hovering flight, for which there is no forward speed, the reference speed U_{ref} is defined as the mean wingtip velocity $2\Phi f R$; the reduced frequency can be re-formed as

$$k = \frac{\pi f c_m}{U_{ref}} = \frac{\pi c_m}{2\Phi R} = \frac{\pi}{\Phi AR},$$ (4.11)

where the AR is introduced here again as in Eq. (4.7). For the special case of 2D hovering airfoils, the reference velocity U_{ref} is the maximum flapping velocity (see Eq. (4.8)), and the reduced frequency is defined as

$$k = \frac{\pi f c_m}{U_{\text{ref}}} = \frac{c_m}{2h_a} = \frac{c}{2h_a}, \tag{4.12}$$

which is simply related to the normalized stroke amplitude. Based on the definition of the reference velocity and reduced frequency, airfoil kinematics Eqs. (4.3) and (4.5) can be rewritten as

$$h(t) = h_a \sin(2kt + \varphi), \tag{4.13}$$

$$\alpha(t) = \alpha_0 + \alpha_a \sin(2kt), \tag{4.14}$$

where t is a dimensionless time, which is nondimensionalized by a reference time $T_{\text{ref}} = L_{\text{ref}}/U_{\text{ref}}$.

In the case of forward flight, another dimensionless parameter is the advance ratio J. In a 2D framework, J is defined as

$$J = U_{\text{ref}}/(2\pi f h_a), \tag{4.15}$$

which is related to St, specifically, $J = 1/(\pi St)$. In Eq. (4.15), the reference velocity U_{ref} is the forward-flying velocity U.

With the reduced frequency, the wing kinematics as illustrated in Eqs. (4.1)–(4.3) can be further re-formed as

$$\phi(t) = \sum_{n=0}^{3} [\phi_{cn} \cos(2knt) + \phi_{sn} \sin(2knt)], \tag{4.16}$$

$$\theta(t) = \sum_{n=0}^{3} [\theta_{cn} \cos(2knt) + \theta_{sn} \sin(2knt)], \tag{4.17}$$

$$\alpha(t) = \sum_{n=0}^{3} [\alpha_{cn} \cos(2knt) + \alpha_{sn} \sin(2knt)], \tag{4.18}$$

where t is a dimensionless time, which is nondimensionalized by a reference time $T_{\text{ref}} = L_{\text{ref}}/U_{\text{ref}}$, resulting in a dimensionless period of π/k.

Finally, if we choose c, U_{ref}, and $1/f$ as the length, velocity, and time scales, respectively, for nondimensionalization, then the corresponding momentum equation for a constant-density fluid yields

$$\frac{k}{\pi} \frac{\partial}{\partial \bar{t}} (\bar{u}_i) + \frac{\partial}{\partial \bar{x}_j}(\bar{u}_j \bar{u}_i) = -\frac{\partial \bar{p}}{\partial \bar{x}_i} + \frac{1}{Re} \frac{\partial^2}{\partial \bar{x}_j^2} (\bar{u}_i), \tag{4.19}$$

where the overbar designates the dimensionless variable. In this form, the reduced frequency and Reynolds number are separated and convenient for investigating their effect.

Information about the fruit fly (*Drosophila melanogaster*), bumblebee (*Bombus terrestris*), hawkmoth (*Manduca sexta*), and hummingbird (*Lampornis clemenciae*) is shown in Table 4.1. For all these flyers, the flapping frequency is around 20-200 Hz,

Table 4.1. *Morphological and flight parameters for selected species*

Parameters	Fruit fly (*Drosophila melanogaster*)	Bumblebee (*Bombus terrestris*) (Dudley and Ellington, 1990a)	Hawkmoth (*Manduca sexta*) (Willmott and Ellington, 1997a, 1997c)	Hummingbird (*Lampornis clemenciae*) (Chai and Millard, 1997; Greenewalt, 1975)
Morphological				
Total mass (body) m (mg)	2.00	175	1579	8400
Wing mass (both wings) m_w (mg)	9.6×10^{-3}	0.9	94	588
Wing length R (mm)	3	13	49	85
Wing area (both wings) S (mm^2)	2.9	106	1782	3524
Wing loading p (N/m^2)	7	16	9	23.5
Wing aspect ratio AR	2.4	6.6	5.3	8.2
Kinematics				
Flapping frequency f (Hz)	200	150	25	23
Stroke amplitude Φ (rad)	2.6	2.1	2.0	2.6
Nominal forward-flight speed U (m/s)	2	4.5	5	8
Reynolds number Re (based on wing chord c)	130–210	1200–3000	4200–5300	11000
Advance ratio J	–	0.66	0.91	0.34

and the flight speed is about several meters per second, yielding a Reynolds number from 10^2 to 10^4 based on the mean chord and the forward-flight speed. In this flight regime, the unsteady effect, inertia, pressure, and viscous forces are all important.

4.2 Nonstationary Airfoil Aerodynamics

Nonstationary airfoil aerodynamics can be largely influenced by vortex dynamics and their subsequent interactions with airfoils. The vortex growth and shedding, if used properly, can provide increased lift generation compared with that of a stationary airfoil. Thus understanding vortex dynamics and its impact on lift generation is of substantial interest to flapping flight.

Hurley (1959) was the first to use vortex flows to enhance high lift at a high AoA. His idea was to exploit, rather than suppress, the LEV. As the incoming flow tends to separate around the leading edge, the boundary-layer control technology is devised to keep the flow attached on the upper surface of a forward-facing flap. A conceptual illustration is shown in Figure 4.5, where a blowing slot is placed at the leading edge of the upper flap, preventing the boundary-layer separation from happening. The LEV that forms above the lower flap enhances the lift, as has been proved in

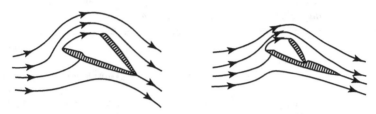

Figure 4.5. Hurley's free-streamline airfoil (Hurley, 1959). Adopted from Wu et al. (1991).

wind-tunnel experiments. After appropriate blowing momentum is chosen, Hurley's airfoil could attain a C_L as high as 5–6 at up to 20°–30° AoA.

A 2D vortex has a tendency to develop spanwise instability and become 3D. Westesson and Clareus (1974) designed an airfoil with special leading and trailing flaps on the upper surface to create a strong axial flow along the spanwise direction to stabilize the vortex pattern. A lift coefficient of $C_L \sim 4.5$ was attained with no blowing at $\alpha = 40°$ (Figure 4.6).

Inspired by the work of Westesson and Clareus (1974), Erickson and Campbell conducted numerous flow-visualization experiments (Erickson and Campbell, 1975) using the same model (see Figure 4.6). They found that a dual corotating vortex system was generated and positioned by blowing, as sketched in Figure 4.6. When the wing leading edge is swept to 45°, the required blowing rate can be significantly reduced by about an order of magnitude. Similar studies have also been done by other researchers, as reviewed by Wu et al. (1991)

Enhanced lift was also reported with the use of the *Kasper wing* (Cox, 1973; Kasper, 1979). As shown in Figure 4.7, the Kasper wing makes use of large-scale anchored vortical flows to create a favorable aerodynamic outcome. It is a successful example of gaining high lift at a poststall AoA in *real flight*, attaining a lift coefficient of 3.15.

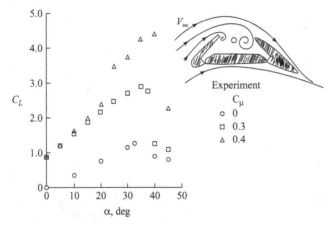

Figure 4.6. Lift enhancement attained by vortex capturing (Erickson and Campbell, 1975). Adopted from Wu et al. (1991).

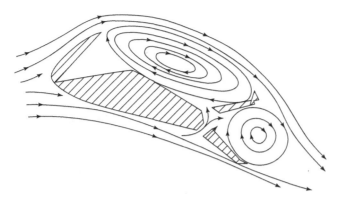

Figure 4.7. Detached vortex on the Kasper wing (Kasper, 1979). Adopted from Wu et al. (1991).

However, wind-tunnel tests have not been successful in reproducing Kasper's flight result, unless external blowing is used (Kruppa, 1977). It is found that, although a lift-enhancing vortex does appear, a counterrotating vortex also accompanies it and largely neutralizes the lift enhancement. It is speculated that, because the real glider wing is not rigid, structurally induced vibration may help enhance the aerodynamics. So far, there has been insufficient information available in the open literature to assess this opinion.

Although biological flyers make use of various physical mechanisms to enhance lift, they share the same characteristics as the previously mentioned design concepts, namely, utilizing large-scale vortical flows for lift enhancement.

4.2.1 *Dynamic Stall*

When an airfoil is accelerated impulsively to constant velocity, the bound vortex needs time to develop to its final, steady-state strength. Depending on the pace of acceleration, it may take up to six chord lengths of travel for the circulation and lift to reach 90% of the final values (Ellington, 1995). However, the fast acceleration of the airfoil can result in lift enhancement that is due to the so-called Wagner effect, which describes the unsteady aerodynamics associated with an accelerating airfoil. Specifically, an impulsively started airfoil develops only a fraction of its steady-state circulation immediately; the steady-state value can be attained only after the airfoil moves through several chord lengths. On the other hand, if the airfoil is started at an AoA above its stalling angle, then a large transient vortex forms above the leading edge, which can dramatically increase the lift (Anders, 2000; Von Karman and Burgers, 1935).

The undulating movements of insect wings make it unlikely that the steady-state value can be reached. Thus, although the quasi-steady estimate of lift is often used in literature, it risks being overly simplistic (Berger, 1999). However, as reported by Dickinson et al. (1999), a lift peak is observed when a wing is accelerated at the beginning of each stroke. Dynamic stall, or delayed stall, is often used to describe

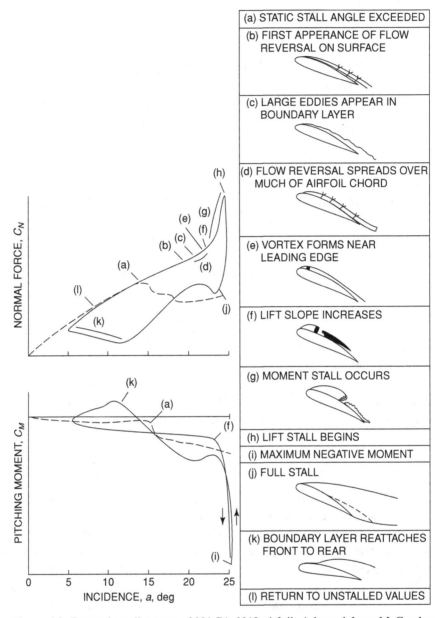

Figure 4.8. Dynamic-stall events of NACA 0012 airfoil. Adopted from McCroskey and Fisher (1972).

the extra lift associated with a wing traveling at high AoAs for a brief period, with a large LEV, before it stalls.

Since the 1950s and 1960s, investigators (Ham, 1968; Harper and Flanigan, 1950; Harris and Pruyn, 1968) have found that the stall does not come instantly when a wing is rapidly pitched beyond the static stall angle. Figure 4.8 depicts the evolution of flow structures in dynamic stall for a rapidly pitching NACA 0012 airfoil (McCroskey and Fisher, 1972). The reverse flow affects the pressure distribution [point (b)] after

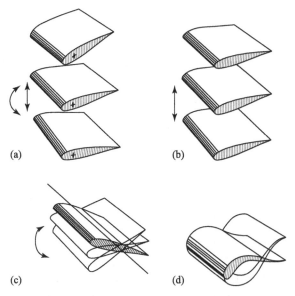

Figure 4.9. Four typical unsteady motions in plunging and pitching motion: (a) combined translational–rotational oscillations, (b) purely translational oscillations, (c) purely rotational oscillations, and (d) advancing wave-type deformations. Adopted from Rozhdestvensky and Ryzhov (2003).

the wing rapidly exceeds the static stall angle (point (a) in Figure 4.8). This reversal progresses up on the airfoil upper surface and forms a vortex. This vortex initially appears near the leading edge of the airfoil (point (e)), enlarges, and then moves down the airfoil. The pitching moment reaches its negative peak and then both lift and pitching momentum start to drop dramatically (points (h) and (i) in Figure 4.8), producing the phenomenon known as dynamic stall. As the AoA decreases, the vortex moves into the wake, and a fully separated flow develops on the airfoil. At the time instant when the AoA reaches its minimum, lift has not reached its minimum value, which indicates that the dynamic stall process forms a hysteresis loop. Figure 4.8 shows such characteristics for the development of lift and pitching momentum. The amplitude and the shape of the hysteresis loop depend on the oscillation amplitude, mean AoA, and reduced frequency.

Most of the research into the dynamic-stall phenomenon has been performed on airfoils oscillating in pitch. This 2D motion has been useful in highlighting the characteristics of dynamic stall on helicopter blades, fish swimming, and flapping flight. The viscous effects play an important role in these cases. This has led to a more careful investigation of the dynamic-stall process, including evaluation of the type of motion (Rozhdestvensky and Ryzhov, 2003) involved (see Figure 4.9). McCroskey et al. (1982) showed the sensitivity to history effects in dynamic stall. They observed that the high-angle part of the oscillating airfoil in a dynamic-stall cycle depends significantly on the rate of change of AoA α near the stall angle; the same lift- and pitching-moment behavior can be attained by matching the rate of change of α at stall limit with different amplitudes.

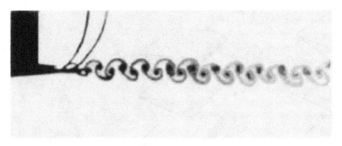

Figure 4.10. Vortices structure behind a stationary NACA 0012 (Lai and Platzer, 1999). (See Plate XIV.)

4.2.2 Thrust Generation of a Pitching/Plunging Airfoil

The first experimental work confirming the possibility of thrust generation on the unsteadily moving wing was done by Katzmayr in 1922 (Katzmayr, 1922). He investigated a fixed wing placed into an oscillating flow field. His studies validated the Knoller–Betz hypothesis (Betz, 1912; Knoller, 1909). Both Knoller and Betz observed that the vertical motion of a flapping wing creates an effective AoA, generating an aerodynamic force with both lift and thrust components. Polonskiy (1948) and Bratt (1953) performed detailed visualizations of the large-scale vortex structures shed from harmonically plunging foils in a uniform flow and observed the characteristics of the vortex structures behind the wing. These experimental observations confirmed the Karman–Burgers thrust-generation hypothesis, i.e., the formation of a reverse Karman vortex street. Polonskiy (1950) and Jones et al. (2001) showed the existence of different types of large-eddy structures generated by oscillating wings in their experiments, and the vortex structures that are shed at an angle to the free stream. Other researchers (Anderson et al., 1998; Freymuth, 1988; Koochesfahani, 1989) studied the 2D flow structure behind oscillating foils and thrust generation, confirming that, depending on the parametric conditions, the wake structure can change from simple sinusoidal perturbations to two or four large-scale eddies. These typical flow structures were captured in visualization experiments by Lai and Platzer (1999). Figure 4.10 shows the typical Karman vortex street behind a stationary NACA 0012, in which clockwise rotating vortices (red) are shed from the upper surface and counterclockwise rotating vortices (green) are shed from the lower surface; Figure 4.11 shows two pairs of vortices shed from the trailing edge per plunge cycle while Figures 4.11(b) and (c) show a single pair with reverse Karman vortex street pattern.

Anderson et al. (1998) experimentally measured the time-averaged thrust coefficient, input power coefficient, and propulsive efficiency of a NACA 0012 airfoil undergoing a combined plunging and pitching motion. They found that the efficiency peaks in the Strouhal number range $0.25 < St < 0.4$, with an efficiency as high as 87% depending on the exact flapping parameters. Satyanarayana and Davis (1978) and Bass et al. (1982) observed the flow structure of an oscillating wing at different Strouhal numbers: For Strouhal numbers [based on the airfoil chord rather than on the plunging amplitude as in Eq. (4.9)] up to 1.0, the shedding of the vortex

(a) $h = 0.0125$ ($kh = 0.098$)

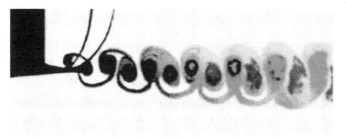

(b) $h = 0.025$ ($kh = 0.196$)

(c) $h = 0.05$ ($kh = 0.393$)

Figure 4.11. Vortex patterns for a NACA 0012 airfoil oscillated in plunge for a free-stream velocity of about 0.2 m/s, a frequency of $f = 2.5$ Hz ($k = 7.85$), and various amplitudes of oscillation (Lai and Platzer, 1999). (See Plate XV.)

street occurs from the sharp trailing edge of the wing; for larger Strouhal numbers, the vortex-shedding location moves from the trailing edge to the lower or upper side of the wing within the period of oscillation.

Another experimental observation discovered the delay of the leading-edge flow separation in unsteady motion. Devnin et al. (1972) indicated that, for a rigid wing with an AR of 1–4 and a NACA 0012 section, no separation is observed up to the instantaneous AoA of $\alpha = 45°$; on the other hand, separation normally occurs at $\alpha = 15°$ ($Re = 10^5$) in the steady case. Taneda investigated the influence of traveling-wave characteristics associated with a flexible plate (Taneda, 1976). He revealed that the turbulence in the boundary layer is suppressed when the speed of propagation of the traveling wave exceeds that of the uniform incoming flow. Researchers have long noticed the interaction of large-scale eddies with oscillating wings (Rosen, 1959) in observing that fish bodies generate large-scale eddies. Using digital particle image velocimetry (DPIV), Mueller (2001) obtained flow patterns behind a swimming fish

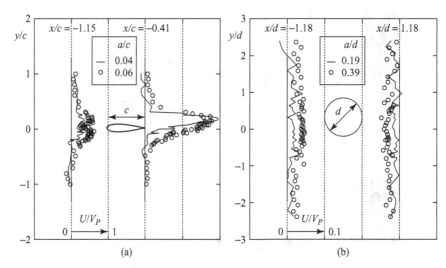

Figure 4.12. Nondimensional mean streamwise velocity profiles generated by (a) a plunging airfoil, and (b) a plunging circular cylinder at $f = 5$ Hz (Lai and Platzer, 2001).

and confirmed that the active vortex control by the body and fin produces a reverse Karman vortex street at maximum propulsive efficiency.

Gopalkrishnan et al. (1994) and Streitlien and Triantaffilou (1998) identified three types of interactions of the harmonically oscillating wing with vortices in the wake: (i) optimal interaction of the new vortices with the vortices shed by the wing, resulting in the generation of more powerful vortices in the reverse Karman vortex street; (ii) destructive interaction of new vortices with those shed by the wing, resulting in the generation of weaker vortices in the reverse Karman street; (iii) interaction of vortex pairs with opposite signs shed from the wing, leading to the generation of a wide wake composed of vortex pairs that are shed at an angle to the free-stream.

In summary, depending on the nature of the interactions between airfoil movement and the associated flow structure, either thrust or drag can be observed. In regard to fluid physics, the flow visualizations of Wolfgang et al. (1998) indicate that the three-dimensionality of the flow is most pronounced near the edges of the body. Furthermore, Triantafyllou et al. (2000) observed that the interaction between large-scale vortex structures generated by the fish body and vortex structures of the fin are important factors in determining the performance of swimming. In the "normal" case, the initial pair of large-scale vortices is generated by the body. Then the body-generated vorticity is redirected by the fin and interacts with the fin-generated vorticity to produce the vortex pair, which is "accurately controlled" by the fish. The timing of vortex formation, propagation, and its instantaneous position are critical for efficient maneuvering and acceleration. Thus the control of vortex generation plays an important role in achieving high locomotion efficiency.

Lai and Platzer (2001) found that, when an airfoil is plunging at zero angle of incidence with no incoming flow, as Figure 4.12(a) shows, a jet is produced by the flapping airfoil and the streamwise velocity downstream of the airfoil is greater than the peak plunge velocity. The jet appears to be biased toward the half-plane above the airfoil. This phenomenon was also observed by Jones et al. (2001). The reason is that, as soon as kh_a (where k is the reduced frequency and h_a is the plunging amplitude normalized by the chord) exceeds approximately 0.8, the vortices shed from the trailing edge are coming too close together and start to interact with each other. Another experiment with a circular cylinder does not observe the jet flow (see Figure 4.12(b)). Therefore it seems that the jet flow is caused by detailed geometry such as curvature and asymmetry of the solid object.

4.3 Simplified Flapping-Wing Aerodynamics Model

A simplified analysis for flapping flight can be established based on the actuator disk model. An actuator is an idealized surface that continuously pushes air and imparts momentum downstream by maintaining a pressure difference across itself, i.e., the lift is equal to the rate of change in fluid momentum. Assuming that insect wings beat at high-enough frequencies so that their stroke planes approximate an actuator disk, the wake downstream of a flapping wing can be modeled as a jet with a uniform velocity distribution (Ellington, 1984a; Hoff, 1919). Although the momentum theory accounts for both axial and rotational changes in the velocities at the disk, it neglects time dependency in wing size, morphology, kinematics, and associated unsteady-lift-producing mechanisms.

By using the Bernoulli equation for steady flow to calculate induced velocity at the actuator disk and the jet velocity in the far wake downstream, i.e., the downwash, Weis-Fogh (1972) derived the induced downwash velocity w_i for a hovering insect at the stroke plane as

$$w_i = \sqrt{\frac{W}{2\pi\rho R^2}}, \tag{4.20}$$

where W is the insect weight, ρ is the air density, and R is the wing length. From the experimental measurements of the beetle *Melolontha vulgaris*, Weis-Fogh (1972) assumed that the downwash velocity in the far wake was twice that at the disk, i.e., $w = 2w_i$, even though he pointed out that w_i varies through a half-stroke and that the stroke-plane amplitude Φ is rarely 180°.

Instead of using a circular disk, Ellington (1984e) proposed a *partial* actuator disk of area $A = \Phi R^2 \cos(\beta)$ that flapping wings cover on the stroke plane, as depicted in Figure 4.13, and modified the expression for the induced power P_{ind}, such that

$$P_{\text{ind}} = \sqrt{\frac{W^3}{2\rho\,\Phi R^2 \cos(\beta)}} = W\sqrt{\frac{1}{2\rho}\left(\frac{W}{A}\right)}, \tag{4.21}$$

where β is the stroke-plane angle and W/A is the disk loading that controls the minimum power requirement. He also noted that, because of the time-varying nature of flapping, a *pulsed* actuator disk seems more representative of hovering flight.

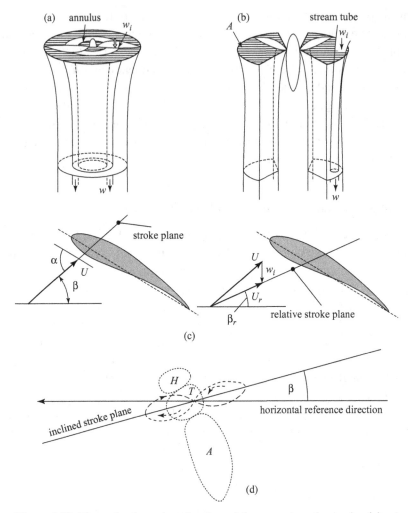

Figure 4.13. The wake flow given by the axial momentum theory for (a) a hovering propeller and (b) a hovering animal. Here w_i is the induced velocity at the disk, w is the vertical velocity attained in the "far" wake, and A is the disk area. (c) Definition of the stroke plane and relative stroke plane. w_i is the is the induced velocity by the vorticity in the wake and β_r is the relative stroke-plane angle. Adopted from Ellington (1984e). (d) Stroke kinematics for a hovering insect; H = head, T = thorax, A = abdomen. Dashed curves represent the upstroke; dotted curves are the downstroke. Adopted from Zbikowski (2002).

He showed that the circulation of the vortex rings in the far wake downstream is related to the jet velocity:

$$\Gamma = \frac{w^2}{2f},\qquad(4.22)$$

where f is the shedding frequency.

Rayner (1979a, 1979b) proposed a method representing the wake of a hovering insect by a chain of small-cored coaxial vortex rings (one produced for each half-stroke). Although the approach could determine the lift and drag coefficients, the effects of stroke amplitude and stroke-plane angle were not accounted for. Sunada and Ellington (2000) developed a method that models the shed vortex sheets in the wake as a grid of small vortex rings with the shape of the grid modeled by wing kinematics so that all forward speeds can be handled.

Overall, the relatively simple approaches just presented are of limited capability because only stroke-plane angle and disk loading are included. The models do not allow, for example, estimation of lift forces for given wing kinematics or wing geometry.

In the quasi-steady approach, the lift and drag coefficients are computed based on the steady-state theory while varying the geometry and speed in time. To account for the variations in velocity and geometry from wing base to wingtip, the blade-element approach has been followed to discretize the wing into chordwise, thin wing strips; the total force is computed by summation of the forces associated with individual strips along the spanwise direction (Ellington, 1984a; Osborne, 1951). Integrating lift over the entire stroke cycle gives the total lift production of the flapping wings. For example, considering such wing kinematics and wing geometry, Osborne (1951) proposed a quasi-steady approach to model insect flight – the forces on the insect wing at any point in time are assumed to be the steady-state values that would be achieved by the wing at the same velocity and AoA. Later, in 1956, Weis-Fogh and Jensen (1956) laid out the basis of momentum and blade-element theories as applicable to insect flight and carried out quantitative analyses on wing motion and energetics available at the time. Their results indicate that, in most cases, when forward flight is considered, the quasi-steady approach appeared to hold for the reason that, as flight velocity increased, unsteady effects diminished. In the mid-1980s, Ellington published a series of papers on insect flight (Ellington, 1984a, 1984b, 1984c, 1984d, 1984e, 1984f). He presented theoretical models for insect flight by using actuator disks (Ellington, 1984e), vortex wake (Ellington, 1984e), quasi-steady methods (Ellington, 1984a), rotation-based mechanisms of clap, peel, and fling (Ellington, 1984e), and insight into unsteady aerodynamics (Ellington, 1984e, 1984f).

From the blade-element method, Ellington combined expressions for lift that is due to translational and rotational phases. Using the thin airfoil theory and the Kutta–Joukowski theorem, he derived the bound circulation as

$$\Gamma_t = \pi c U \sin \alpha, \tag{4.23}$$

where c is the chord length, U is the incident velocity, and α is the *effective* AoA corrected for profile shape. Following Fung's method (Fung, 1969), he also derived an expression for circulation that is due to rotational motion by computing incident velocity at the 3/4 chord point while satisfying the Kutta–Joukowski condition, giving

$$\Gamma_r = \pi \dot{\alpha} c^2 \left(\frac{3}{4} - \hat{x}_0 \right), \tag{4.24}$$

where $\dot{\alpha}$ is the rotational (pitching) angular velocity and $\hat{x}_0$ is the distance from the leading edge to the point about which rotation is being made (pitch axis), normalized with respect to the chord c. Combining the preceding two expressions, Ellington obtained the quasi-steady lift coefficient:

$$C_L = 2\pi \left[\sin\alpha + \frac{\dot{\alpha}c}{U} \left(\frac{3}{4} - \hat{x}_0 \right) \right]. \tag{4.25}$$

Furthermore, to determine lift and power requirements for hovering flight, Ellington (1984f) sought estimates for the mean lift coefficient through the flapping cycle and derived a nondimensional-parameter-based expression:

$$\bar{C}_L = \frac{8\bar{L}\cos^2(\beta_r)}{\rho f^2 \Phi^2 R^2 \hat{r}_2^2(S)\overline{\left(d\hat{\phi}/d\hat{t}\right)^2} S \cos^2(\beta)}, \tag{4.26}$$

where $\bar{L}$ is the mean lift through a <u>half-stroke</u>, ρ is the air density, f is the wing-beat frequency, Φ is the stroke angle, $\overline{(d\hat{\phi}/d\hat{t})^2}$ is the mean-squared flapping angular velocity, S is the wing area, β is the stroke-plane angle, β_r is the relative stroke-plane angle (see Figure 4.13(c)), and r_2 is the second moment of wing area.

Numerous versions of the quasi-steady approach can be found in the literature; in general, the model predictions are not consistent with the physical measurements, especially when the hovering flight of insects is considered. For example, lift coefficients obtained under those conditions yield (i) 0.93–1.15 for a dragonfly, *Aeschna juncea* (Newman et al., 1977; Wakeling and Ellington, 1997b), (ii) 0.7–0.78 for a fruit fly, *Drosophila* (Vogel, 1967; Zanker and Gotz, 1990), and (iii) 0.69 for a bumblebee, *Bombus terrestris* (Dudley and Ellington, 1990b). However, lift coefficients estimated by direct force measurements in flying insects are significantly larger than those predicted by the quasi-steady methods, ranging from 1.2 to 4 for various insects including the hawkmoth, *Manduca sexta*, bumblebee, *Bombus terrestris*, parasitic wasp, *Encarsia formosa*, dragonfly, *Aeschna juncea*, and fruit fly, *Drosophila melanogaster* (Ellington, 1984e; Lehmann and Dickinson, 1998; Norberg, 1975; Weis-Fogh, 1972).

As quasi-steady methods are unable to predict accurately flapping-wing aerodynamics, empirical corrections have been introduced. Walker and Westneat (2000) presented a semiempirical model for insectlike flapping flight, which includes, e.g., Wagner's function (Fung, 1969), which is devised to account for the lift enhancement caused by an impulsively starting airfoil. They used a blade-element method to discretize the flapping wing and compute forces on the wing elements, in which the forces comprise a circulation-based component and a noncirculatory apparent mass contribution. Sane and Dickinson (2001) refined a quasi-steady model to describe the forces measured in their earlier experiments on the *Robofly* (see Figure 4.14), a mechanical, scaled-up model of the fruit fly, *Drosophila melanogaster* (Fry et al., 2003). They decomposed the total force F into four components, namely

$$F = F_t + F_r + F_a + F_w, \tag{4.27}$$

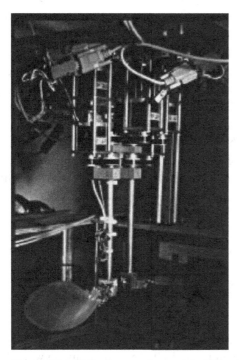

Figure 4.14. *Robofly* experimental facilities developed by Lehmann, Dickinson, and others (http://www.biofuture-wettbewerb.de/index.php?index=18).

where the subscripts t and r are for translational and rotational quasi-steady components, respectively, a is for added mass (added mass arises because an accelerating or decelerating body must move some volume of surrounding fluid as it moves through it), and w is for wake capture. In the blade-element approach, a *Robofly* wing is divided along the spanwise direction into chordwise strips, and the forces on each strip are computed individually and integrated along the span. The translational quasi-steady force F_t is computed from empirically fitted equations. To determine the rotational quasi-steady force F_r, Sane and Dickinson (2001) set the forces that are due to added mass and wake capture to zero, i.e., $F_a = F_w = 0$, by removing any accelerations and avoiding wake reentry, respectively. They measured the force F acting on a wing undergoing a constant translation and rotation for one forward stroke only and obtained the rotational force F_r by subtracting the empirically predicted translational force F_t from the measured total force F. Furthermore, an analytical method was developed to compute the added mass F_a; and, knowing these three components, they evaluated the wake-capture force F_w by subtracting the components F_t, F_r, and F_a from the total measured force F.

Recently a model for unsteady lift generation for insectlike flapping wings was proposed by Pendersen and Zbikowski (2006). The model is modular, giving a better insight into various effects on aerodynamic force generation, and includes added-mass effects, the quasi-steady assumption, a LEV effect, and the wake effect. The model's predicted lift and drag forces were compared with the measurements of

Dickinson et al. (1999), and, despite its simplifications, the model captures reasonably well the lift evolution, but overpredicts the force peak values.

Although such semiempirical methods can be tuned to provide good agreement with experimental measurements, their predictable capabilities are questionable because they cannot adequately account for the relevant unsteady, vortical fluid physics. In particular, the unsteady effects are important not only during the translational phases of the stroke (upstroke and downstroke) but also during the rotational phase near the end of each stroke when the wings are quickly rotated around their spanwise axes (Dickinson and Gotz, 1993; Liu et al., 1998; Wakeling and Ellington, 1997a, 1997b). Nevertheless, the quasi-steady model does provide some insight into flapping flight in insects and birds and offers quick estimates of unsteady aerodynamic coefficients.

Next, we review the major physical mechanisms responsible for lift enhancement of flapping wings and a series of case studies aimed at elucidating these mechanisms.

4.4   Lift-Enhancement Mechanisms in Flapping Wings

Robotic models, illustrated in Figure 4.14, offer a valuable framework for studying biological flight. As previously discussed, with geometric and kinematic similarities, one can maintain the dynamic similarity by scaling up the wing dimension while appropriately lowering the flapping frequency, rendering the Reynolds number or the reduced frequency unchanged. Ellington et al. (1996) used a robotic model to investigate flow over the wings of a hovering hawkmoth and discovered that the LEV spiraled out toward the wingtip. Their finding provided a qualitative explanation of one particular high-lift mechanism. Dickinson et al. (1999) also used a robotic model, representing a fruit fly, to directly measure forces and visualize the flow patterns around a flapping wing. They demonstrated two force peaks at the rotation phase, namely, the rotational mechanism associated with fast pitch-up, and the wake-capture mechanism resulting from the airfoil and vortical flow interactions. Although different explanations of the two force-generation mechanisms have been offered, as described in the following subsection, it is clear that a robotic model offers better control and improved experimental resolution in studying flapping-wing flight.

Birch and Dickinson (2001) further observed substantially different flow patterns around the same model, based on large moths and small flies, to investigate the impact of the scaling parameters on the aerodynamic outcome. Further refining the experimental techniques, Fry et al. (2003) recorded, with a 3D infrared high-speed video, the wing and body kinematics of free-flying fruit flies performing rapid flight maneuvers, and "replayed" them on their robotic model to measure the aerodynamic forces produced by the wings. They reported that the fly generated sufficient torque for rapid turn with subtle modifications in wing motion, and suggested that inertia, not viscous force, dominates the flight dynamics of flies.

To date, these robotic-model investigations have focused on the flapping pattern of rigid wings without accounting for structural flexibility. Combes and Daniel (2003) analyzed the bending and flexion of the wings of a hawkmoth's flapping flight. They reported that, compared with the inertial forces, the aerodynamic forces play a minor role in determining wing deformation during flapping movement.

In addition to robotic models, high-speed measurements for real flyers were also conducted. For example, Srygley and Thomas (2002) reported a study on the force-generation mechanisms of free-flying butterflies by using high-speed, smoke-wire flow visualizations to obtain qualitative images of the airflow around flapping wings. They observed clear evidence of LEV structures. In comparison, in moth and fly flight, the helical structure and the spanwise, axial flow patterns appear to be much weaker. They suggest that free-flying butterflies use a variety of aerodynamic mechanisms to generate force, including wake capture, LEVs, active and inactive upstrokes, rapid rotation, and clap-and-fling; these different mechanisms are often used in successive strokes as seen during take-off, maneuvering, maintaining steady flight, and landing.

Warrick et al. (2005) used DPIV to observe the wake around hovering hummingbirds. They observed force asymmetry between the upstroke and the downstroke. Specifically, 75% of the lift is generated during the downstroke and 25% during the upstroke. They reported inversion of the cambered wings during the upstroke, as well as evidence of LEVs, created during the downstroke. As suggested by the Reynolds number, a hummingbird's aerodynamics regime overlaps that of larger insects.

Videler et al. (2004) recorded a water-tunnel experiment in which they used the DPIV technique for flow around a single wing of a swift in fast gliding. Their results show that gliding swifts can generate stable LEVs at small (5°–10°) AoAs. Whereas the swept-back hand-wings generate lift with LEVs, the flow around the arm-wings seems to remain attached.

Clearly, depending on the size and flow parameters of individual species, various lift-enhancement mechanisms are observed. For example, the delayed-stall phenomenon has been investigated from both dynamic-stall (Dickinson et al., 1999; Lehmann et al., 2005) and upper-wing LEV (Ellington et al., 1996; Maxworthy, 1979) viewpoints. The high-lift peak during wing pitch-up has been explained by use of the Magnus effect (Lehmann et al., 2005) (the lift generated by a rotating object by means of the induced velocity differential between upper and lower surfaces), as well as vortical flow structures (Ellington et al., 1996). As already mentioned, a high-lift peak after a wing reverses its direction can result from wake-capture and/or fast acceleration processes. The wake capture produces aerodynamic lift by a transfer of fluid momentum associated with large-scale vortical flow shed from the previous stroke to the wing at the beginning of each half-stroke. The fast acceleration results in lift enhancement and can be explained in part by the so-called Wagner effect discussed in Subsection 4.2.1.

The preceding discussion offers a sample of the experimental and modeling investigations. In the following subsection, we address flapping-wing aerodynamics by focusing on specific unsteady-lift mechanisms, as well as related scaling, geometric, and kinematic parameters. Overall, four major lift-enhancement mechanisms associated with flapping-wing aerodynamics have been reported in the literature:

1. delayed stall that is due to LEVs,
2. aerodynamic peak that is due to pitch-up rotation,
3. wake capture that is due to vortical flow and airfoil interactions,
4. Weis-Fogh's clap-and-fling dynamics.

These mechanisms and their impact on aerodynamics are discussed in the following subsections.

4.4.1 *Leading-Edge Vortex*

In addition to the aerodynamics literature reviewed earlier in this chapter, other researchers have long recognized the potential benefit of trapped or wing-attached vortices in flapping-wing lift enhancement (Bradley et al., 1974; Campbell, 1976; Dickinson and Gotz, 1993; Maxworthy, 1979; Sunada et al., 1993). In particular, the high-lift mechanism generated by the LEV in a flying insect has received substantial attention, following the original discovered by Ellington et al. (1996). It appears that the LEV can enhance lift by attaching the bounded vortex core to the upper leading edge during wing translation (Ellington et al., 1996; Houghton and Carpenter, 2003; Usherwood and Ellington, 2002; Van den Berg and Ellington, 1997). The LEVs generate a lower-pressure area, which results in a large suction on the upper surface. It seems that the lift enhancement can sustain 3 or 4 chord lengths of travel before vortex breakdown occurs.

Ellington and coworkers designed a 10:1 scaled-up, robotic model previously discussed to study the hawkmoth, *Manduca sexta*. To maintain both the Reynolds number and the reduced frequency similarity in hovering, as introduced in Subsection 4.1.2, they preserve fR^2 between the real insect and the mechanical model, where f is flapping frequency and R is wing length. The robotic model is approximately 10 times larger than the hawkmoth and accordingly flaps its wings in air at a frequency of 0.3 Hz. Using geometrically similar hawkmoth wing models that undergo hovering with the same flapping kinematics can therefore satisfy the aerodynamic similarity. By using smoke streams to visualize the flow around a flapping wing, Ellington et al. (1996) demonstrated the presence of a vortex close to the leading edge of the wing. They observed a small but strong LEV that persists through each half-stroke (down-stroke). From direct observation, they proposed that the LEV is responsible for the augmented lift forces. The LEV has a high axial flow velocity in the core and is stable, separating somewhat from the wing at approximately 75% of the wing length span-wise and then connecting to a large, tangled tip vortex. The overall vortical structures are qualitatively similar to those of low-AR delta wings (Ellington et al., 1996; Van den Berg and Ellington, 1997) that stabilize the LEV by maintaining the spanwise pressure gradient, increasing lift well above the critical AoA. They have further suggested that the vortex stability in flapping wings is maintained by a spanwise axial

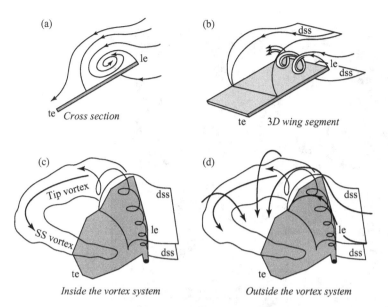

Figure 4.15. Spatial flow structure of LEVs: le designates leading edge, te designates trailing edge, dss designates dividing stream surface, SS vortex designates combined starting/stopping vortex. Adopted from Van den Berg and Ellington (1997).

flow along the vortex core (see Figure 4.15), creating "delayed stall," to enhance lift during the translational phase.

Liu and Kawachi (1998) and Liu et al. (1998) conducted unsteady Navier–Stokes simulations of the flow around a wing of a hawkmoth, *Manduca sexta*, to probe the unsteady aerodynamics of hovering flight. They adopted a realistic geometric wing model as well as flapping kinematics and presented the salient features of the LEV and the spiral axial flow during translational motions. Their results are consistent with those observed by Ellington et al. (1996). Figure 4.16 shows that (i) the LEV created during previous translational motion and (ii) the vortical flows established during the rotational motions of pronation and supination together form a complex flow structure. They estimated that lift is produced mainly during the downstroke and the latter half of the upstroke, with little force generated during pronation and supination.

Dickinson and Götz (1993) measured the aerodynamic forces of an airfoil impulsively started at high AoAs in the Reynolds number range of the fruit fly wing ($Re = 75$–225). They observed that, at AoAs above $13.5°$, impulsive movement resulted in the production of a LEV that stayed attached to the wing for the first two chord lengths of travel, resulting in an 80% increase in lift compared with the performance measured five chord lengths later.

The LEV as a lift-enhancement mechanism has been questioned by Zbikowski (Zbikowski, 2002) because a dynamic-stall vortex on an airfoil is found to break away and convect elsewhere as soon as the wing translates (McCroskey et al., 1982). Nevertheless, LEVs have been observed on the wings of insects, as well as on robotic

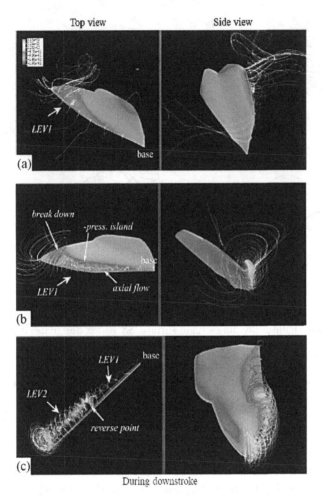

Figure 4.16. Wing-surface pressure and streamlines revealing the vortical structures for the 3D numerical simulation of a hovering hawkmoth (Liu et al., 1998): (a) positional angle $\phi = 30°$; (b) $\phi = 0°$; (c) $\phi = -36°$. The Reynolds number is approximately 4000 and the reduced frequency k is 0.37. (See Plate XVI.)

models (Birch and Dickinson, 2001; Ellington et al., 1996; Liu et al., 1998; Van den Berg and Ellington, 1997). Usherwood and Ellington (2002) showed that axial flow is the predominant factor in LEV stabilization. Specifically, they found that the LEV is generated as soon as the wing starts to revolve, resulting in maximum lift coefficients well above the corresponding 2D steady-state values. Hence the LEV is essentially anchored on a wing's upper surface while it flaps. It should be noted that, although helicopter blade models have been used to help explain flapping-wing aerodynamics, spanwise axial flows are generally considered to play a minor role in influencing helicopter aerodynamics (De Vries, 1983; McCroskey et al., 1976). In particular, helicopter blades operate at a substantially higher Reynolds number and a lower AoA. The much larger AR of a blade also makes the LEV harder to anchor.

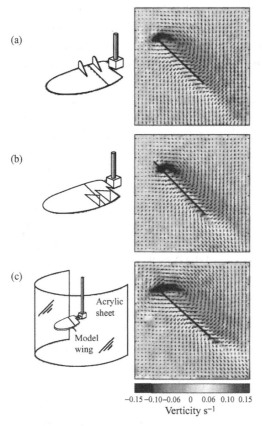

Figure 4.17. The LEV still exists on the wing despite manipulations of the spanwise flow by (a) forward-pointing fences; (b) backward-pointing fences; (c) wall. The wall prevents the LEV from extending farther along the wing before detaching into a tip vortex. Adopted from Birch and Dickinson (2001).

Birch and Dickinson (2001) investigated the LEV features of the flapping wings of the fruit fly at a Reynolds number of 160. They reported that, in contrast to the hawkmoth LEV, which detaches from the wing surface at approximately 75% of the wing length with the presence of a strong axial flow in the core, the LEV of the fruit fly exhibits a stable vortex structure without separation during most of the translational phases. Furthermore, there is little axial flow in the vortex core, amounting to only 2% to 5% of the averaged tip velocity (Figure 4.17). However, strong spanwise flow is observed at the rear two-thirds of the chord, at about 40% of the wingtip velocity. For a fruit fly, the LEV is observed to be stably attached throughout the half-stroke without breaking up. Observing the considerable difference exhibited between fruit fly and hawkmoth models, Birch and Dickinson (2001) hypothesized that the attenuating effect of the downwash induced by the tip vortex and wake vorticity limits the growth of the LEV by lowering the effective AoA and prolonging the attachment of the LEV.

Another recent study (Srygley and Thomas, 2002) on large red admiral butterflies, *Vanessa Atlanta*, also questioned the existence of axial flow even at the level of the

Reynolds numbers comparable with that of hawkmoths. They used smoke trails to visualize the wake about free-flying butterflies in a wind tunnel and showed that the LEV spreads from the wing surface to the body of the animal. In contrast to the conical LEV observed in the hawkmoth, the butterfly LEV exhibits a more cylindrical-shaped vortex with constant diameter and at the end connects with the tip vortex. Because the helical structure of the LEV is much weaker on a butterfly wing, the general role of axial flow for stabilizing the LEV is again questionable.

Thomas et al. (2004) showed, by qualitative free- and tethered-flight flow visualization, that dragonflies attain lift by generating high-lift LEVs. Specifically, in normal free flight, dragonflies use counterstroking kinematics, with an LEV on the forewing downstroke, attached flow on the forewing upstroke, and attached flow on the hindwing throughout. On the other hand, accelerating dragonflies switch to in-phase wing beats with highly separated downstroke flows, with a single LEV attached across both the forewings and hindwings. Also, the flow visualizations suggested that spanwise flow is not a dominant feature of the flow field, as it has been observed that spanwise flows sometimes run from wingtip to centerline, or vice versa, depending on the degree of sideslip. LEV formation always coincides with rapid increases in the AoA, and the smoke visualizations clearly show the formation of LEVs whenever a rapid increase in the AoA occurs. Furthermore, they think that the flow fields produced by dragonflies differ qualitatively from those published for mechanical models of dragonflies, fruit flies, and hawkmoths, which preclude natural wing interactions. However, controlled parametric experiments show that, provided the Strouhal number is appropriate and the natural interaction between left-hand and right-hand wings can occur, even a simple plunging plate can reproduce the detailed features of the flow seen in dragonflies. They suggest that stability of the LEV is achieved by a general mechanism whereby flapping kinematics is configured so that an LEV would be expected to form naturally over the wing and remain attached for the duration of the stroke.

Using 3D Navier–Stokes computations, Viieru et al. (2006) and Shyy and Liu (2007) investigated the Reynolds number effect on the LEV for hovering flight. It is found that the LEV structures are strongly affected by the Reynolds number (defined in Eq. (4.7)). Figure 4.18 shows the streamline patterns at three Reynolds numbers; Figure 4.18(a) corresponds to a hawkmoth hovering at $Re_{f3} = 6000$, Figure 4.18(b) corresponds to a fruit fly hovering at $Re_{f3} = 120$, and Figure 4.18(c) corresponds to a thrips hovering at $Re_{f3} = 10$. At $Re_{f3} = 6000$, as observed experimentally (Ellington et al., 1996), an intense, conical LEV core is observed on the paired wings with a substantial spanwise flow at the vortex core, breaking down at approximately three-quarters of the span toward the tip. At $Re_{f3} = 120$ [Figure 4.18(b)], the vortex no longer breaks down and is connected to the tip vortex. The spanwise flow at the vortex core becomes weaker as the Reynolds number is lowered, which is in qualitative agreement with the findings of Birch and Dickinson (2001). When the Reynolds number is reduced further to $Re_{f3} = 10$, a vortex ring connecting the LEV, the tip vortex, and the trailing vortex is observed (Figure 4.18(c)); the flow structure shows more of a cylindrical than a conical form. Inspecting the momentum equation, one can see that

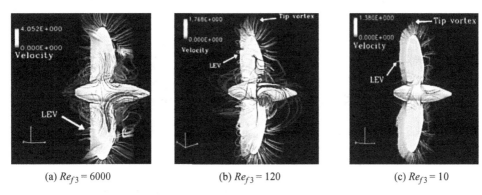

(a) $Re_{f3} = 6000$ (b) $Re_{f3} = 120$ (c) $Re_{f3} = 10$

Figure 4.18. Numerical results of leading-edge vortical structures at different Reynolds numbers. (See Plate XVII.)

the pressure gradient, the centrifugal force, and the Coriolis force together are likely to be responsible for the LEV stability.

To identify the roles of the translational and rotational motions of a flapping wing in the formation of the LEV, computed velocity-vector distributions on an end-view plane, at 60% of wingspan for $Re_{f3} = 6000$ (hawkmoth) are compared against those for $Re_{f3} = 134$ (fruit fly) in Figures 4.19(a) and 4.19(b). The influence of wing rotation on the LEV is more evident at the lower Reynolds number (134) than at the higher one (6000). On the other hand, the higher Reynolds number (6000) yields a much more pronounced axial flow at the core of the LEV, which together with the LEV forms a helical flow structure near the leading edge. In contrast, only very weak axial flow is detected for the lower Reynolds number (134). Figures 4.19(c) and 4.19(d) illustrate the pressure-gradient contours on the wing of a fruit fly model and a hawkmoth model, respectively. Compared with hawkmoths, fruit flies, at a Reynolds number of 100–250, cannot create as steep pressure gradient at the vortex core; nevertheless, they seem to be able to maintain a stable LEV during most of the downstroke and upstroke. Whereas the LEV on a hovering hawkmoth's wing breaks down in the middle of the downstroke, the LEV on the hovering fruit fly's wing stays attached during the entire downstroke, eventually breaking down during the subsequent supination.

Birch et al. (2004) conducted flow visualization around a robotic fruit fly model wing, and also noticed that, although the LEV remains stable at both lower ($Re_{f3} = 120$) and higher ($Re_{f3} = 1400$) Reynolds numbers, the flow changes from a relatively simple pattern at a lower Reynolds number to spiral flow at a higher Reynolds number. Vorticity measurements taken at midstroke, in a plane located at 0.65 of the wing length R and perpendicular to the spanwise direction, show a stronger and larger LEV for the higher Reynolds number [Figure 4.20(d)] associated with intense axial (spanwise) velocity within the LEV core, with magnitudes significantly larger then those of the tip velocity (Birch et al., 2004). At a lower Reynolds number (120), no peak in axial flow has been observed in the area of the LEV core (Figure 4.20(c)), likely because of the stronger viscous effect.

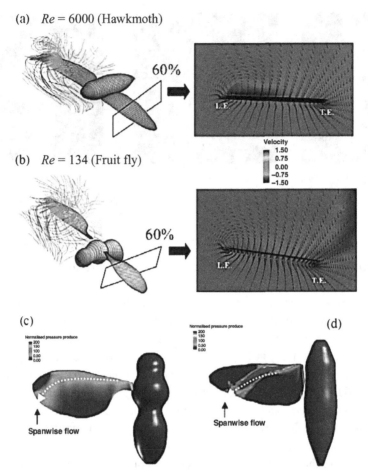

Figure 4.19. Comparison of near-field flow fields between a fruit fly and a hawkmoth. Wing-body computational model of (a) a hawkmoth ($Re_{f3} = 6000$, $U_{ref} = 5.05$ m/s, $c_m = 1.83$ cm), and (b) a fruit fly model ($Re_{f3} = 134$, $U_{ref} = 2.54$ m/s, $c_m = 0.78$ mm), with the LEVs visualized by instantaneous streamlines and the corresponding velocity vectors in a plane cutting through the left wing at 60% of the wing length. Pressure-gradient contours on the wing surface for (c) a fruit fly, and (d) a hawkmoth. The pressure gradient indicates the direction of the spanwise flow. (See Plate XVIII.)

The LEV of a flapping wing plays a role similar to that of a fixed delta wing. The delta wing owes much of the lift that it is able to generate to the fact that the vortex flow initiates at the leading edge of the wing and rolls into a large vortex over the leeward side, containing a substantial axial velocity component. This high-flow velocity in the core of the vortex is a region of low pressure, which generates a suction, i.e., lift. For a delta wing placed at high AoAs, vortex breakdown occurs, causing the destruction of the tight and coherent vortex. The diameter of the core increases, and the axial velocity component is no longer unidirectional. With the loss of axial velocity the pressure increases, and consequently the wing loses lift. The literature on the subject is immense, and for a more general presentation of the various aspects of vortex breakdown, we refer the reader to several review articles (Escudier, 1988; Hall, 1972; Leibovich, 1978). For a fixed wing, an important trend is that, at a fixed

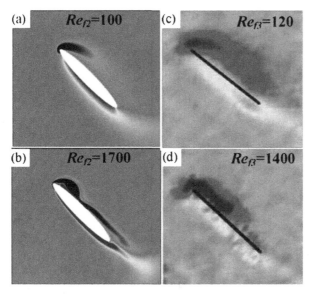

Figure 4.20. Vortical flow structures for pitch-up airfoils: (a), and (b) computational results for flow over a 2D elliptic airfoil undergoing water-treading hovering at two Reynolds numbers. The airfoil position corresponds to the midstroke, where the pitch angle reaches the maximum value. (c), (d) Experimental vorticity field side views for a fruit fly modeled wing at $0.65R$ at midstroke. The experimental information in (c) and (d) is reprinted from Birch et al. (2004). (See Plate XIX.)

AoA, if the swirl is strengthened, then vortex breakdown occurs at lower Reynolds numbers. On the other hand, a weaker swirling flow tends to break down at a higher Reynolds number. Because the fruit fly exhibits a weaker LEV, from this viewpoint, it tends to maintain the vortex stracture better than a hawkmoth, which creates a stronger LEV. Of course, the link regarding vortex breakdown between a fixed and a flapping wing, if any, is not established.

It is noted that 2D flow simulations can also yield features similar to those just discussed. For example, for an elliptic airfoil following the water-treading hovering mode (illustrated in Figure 4.3(a)), noticeable effects of the Reynolds number are observed. Consider the case in which, at the midstroke, the airfoil reaches the maximum pitch-up angle and translational velocity. At a lower Reynolds number, $Re_{f2} = 100$, the vortical flow in the leading-edge region is weaker and less capable of making the turn to stay close to the solid surface (Figure 4.20(a)) than that of a higher Reynolds number, $Re_{f2} = 1700$, case (as seen in Figure 4.20(b)).

It seems that the leading-edge vortical flow structures are influenced by the interplay between the swirl strength and the Reynolds number, as well as the flapping kinematics such as rotational rates. Further investigations are needed to better understand the role of the wing shape in light of the unsteady, large scale vortical flow structures associated with flapping wings.

4.4.2 Rapid Pitch-Up

The LEV-based lift-enhancement mechanism seems to be a main feature during the translational motion of the stroke. On the other hand, the flapping wings

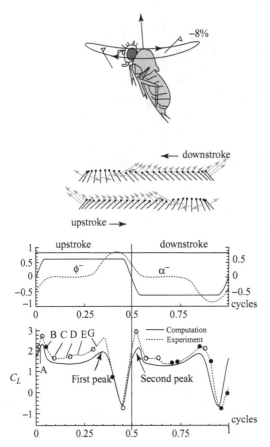

Figure 4.21. Experimental and numerical lift coefficients for a fruit fly-modeled wing, showing the two lift peaks at the end of the upstroke and the beginning of the downstroke. The first lift peak is associated with the rapid vorticity increase in which the wing performs advanced rotation at the end of the stroke. The Reynolds number is 136. Adopted from Sun and Tang (2002a).

also experience rapid wing rotation at the ends of the downstroke and upstroke, which can enhance the lift force in flying insects.

Kramer (1932) first demonstrated that a wing can experience lift coefficients above the steady-stall value when the wing is rotating from low to high AoAs, which is termed the Kramer effect. The unsteady aerodynamic characteristics associated with the time-dependent AoA, including hysteresis, were illustrated in Figure 4.8. Dickinson et al. (1999) used their *Robofly* (see Figure 4.14) along with varied rotational patterns, illustrated in Figure 4.4, to investigate the interplay between kinematics and lift generation. They identified two aerodynamic force peaks at the end and the beginning of each stroke (pronation and supination). The first force peak can be explained based on the rotational circulation. The resulting force enhancement is influenced by the timing of wing rotation while translating. They found that an advanced rotation produces a mean lift coefficient of $C_L = 1.74$, almost 1.7 times higher than that of a delayed rotation ($C_L = 1.01$); a symmetrical rotation can attain

(a) $\tau = 16.7, \alpha = 134.4°$ (b) $\tau = 17.5, \alpha = 140.0°$ (c) $\tau = 18.0, \alpha = 140.0°$ (d) $\tau = 18.8, \alpha = 140.0°$

(e) $\tau = 19.4, \alpha = 140.0°$ (f) $\tau = 20.6, \alpha = 122.4°$ (g) $\tau = 20.8, \alpha = 113.3°$ (h) $\tau = 22.0, \alpha = 52°$

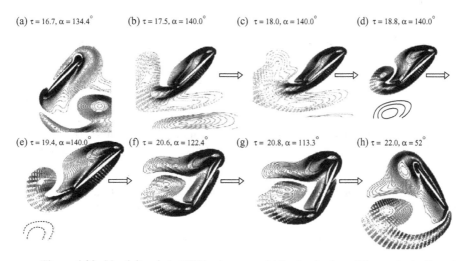

Figure 4.22. Vorticity plot at 75% wingspan. At the beginning of the upstroke the wing accelerates with little changes in the AoA [from $\alpha = 46°$ at (a) to $\alpha = 40°$ at (c)]. The fast acceleration increases the rate of change of fluid momentum, and an increase in lift is observed. From (c) to (e) the wing translates with constant speed at $\alpha \sim 40°$. The dynamic-stall vortex does not shed, and a large lift can be maintained. From (e) to (f) the wing pitches-up rapidly while translating with constant speed, resulting in a sharp increase of lift and drag (first peak in Figure 4.21). From (g) to (h) the wing decelerates and the lift is decreasing quickly. The Reynolds number is 136. Adopted from Sun and Tang (2002a).

a value of $C_L = 1.67$. These peaks were confirmed by the numerical simulations of Sun and Tang (2002a) and Ramamurti and Sandberg (2001). In addition, Sun and Tang (2002b) further investigated three mechanisms responsible for lift enhancement by means of unsteady aerodynamics, namely, (i) rapid acceleration of the wing at the beginning of a stroke, (ii) delayed stall, and (iii) fast pitch-up rotation of the wing near the end of the stroke.

As shown in Figure 4.21, the first peak, termed "rotational force" by Sane and Dickinson (2002), appears near the end of each stroke. In advanced rotation, the wing flips before reversing its translational direction, as illustrated in Figure 4.22, and the leading edge rotates backward relative to the translation. From their computational analysis, Sun and Tang (2002a) suggest that the first peak is due to a rapid vorticity increase when the wing experiences fast pitch-up rotation. The pitch-up rotation and the associated vorticity increase are plotted in Figures 4.22(f) and 4.22(g). Sane and Dickinson (2002) attributed this first force peak to the additional circulation generated to reestablish the Kutta condition during rotation. Overall, the findings reported by Sun and Tang (2002a) and Sane and Dickinson (2002) are in agreement. The second peak, termed wake capture, is related to the wing–wake interaction and is discussed next. Together, these two peaks contribute to lift enhancement.

Because both pitch-up and wake capture are strongly influenced by flapping kinematics, more discussion is offered later to help elucidate the parametric variations of these factors.

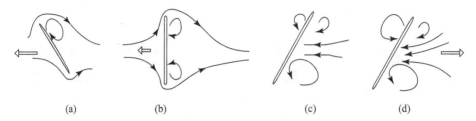

(a)   (b)   (c)   (d)

Figure 4.23. Momentum transfer in a wake-capture interaction: (a) wing is steadily translating; (b) trailing-edge vortex is generated as the wing rotates around a spanwise axis; (c) LEVs generated when the wing is rotating at a very high flapping speed; (d) wing reverses flapping direction and encounters the induced velocity field and a fluid momentum is transferred to the wing that generates a peak in the aerodynamic force.

4.4.3 Wake Capture

As discussed by Dickinson et al. (1999), the wing–wake interaction can significantly contribute to lift production in hovering insects. In Figure 4.21, the second peak is generated at the beginning of each stroke of hovering flight when the wings reverse the direction of moving while rotating about the spanwise direction. The physical mechanism, termed wake capture, produces aerodynamic lift by a transfer of fluid momentum to the wing at the beginning of each half-stroke. A wing meets the wake created during the previous stroke after reversing its direction, thus increasing the effective flow speed surrouding the airfoil, which generates the second force peak. The schematic of the wake-capture mechanism is illustrated in Figure 4.23.

The effectiveness of wake capture is a function of wing kinematics and flow structure. It should be noted that this second force peak is apparently distinct from rotational lift because its timing is independent of the phase of wing rotation. Dickinson et al. (1999) showed that the second peak persists even by halting the wing after it rotates, indicating that the wake produced by the wing motion in the previous half-stroke serves as an energy source for lift production. This is illustrated in Figures 4.24, 4.22(a), and 4.22(b).

The wake-capture mechanism during hovering flight was also investigated by Viieru et al. (2006) and Tang et al. (2007). They investigated flows over a hovering 2D elliptic airfoil with 15% thickness by using two kinematic patterns, one termed the water-treading mode (Figure 4.3(a)) and the other the normal hovering mode (Figure 4.3(b)). Both modes are characterized by the same sinusoidal variation with a stroke amplitude of $h_a = 1.4c$ (Eq. (4.4)) and a pitch angle amplitude of $\alpha_a = 45°$ (Eq. (4.5)); the only difference is the initial pitch angle. Given the stroke amplitude, the reduced frequency, defined by Eq. (4.12), is $k = 0.357$ and the Reynolds number is $Re_{f2} = 100$.

The sinusoidal motion along a horizontal stroke plane is similar to that given by Wang et al. (2004), who conducted a 2D simulation of a hovering elliptic airfoil with the stroke amplitude h_a between $1.4c$ and $2.4c$, leading to a reduced frequency (Eq. (4.12)) k between 0.36 and 0.21. The Reynolds number considered is between 75 and 115.

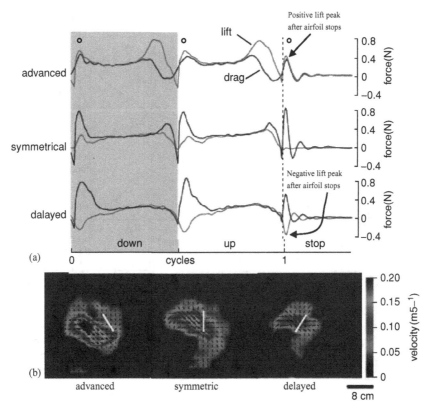

Figure 4.24. Evidence of wake-capture mechanism and its effect on force generation for a robotic fruit fly-modeled wing at $Re_{f3} = 136$: (a) Lift and drag forces for a full flapping cycle followed by a complete stop at the end of the upstroke for different phases of wing rotation; (b) DPIV-generated flow images showing the flow around the midchord of the wing. The fluid velocity orientation is indicated by the arrows and the magnitude by the arrows' length and background colors. The rotational circulation generated and shed from the previous stroke transfers momentum to the wing even after a complete stop. The flow patterns are similar between the different wing-rotation phases; however, the fluid velocities are greater when the rotation is advanced. The strong incoming flow for advanced rotation combined with the wing position generates a positive lift peak after the wing is stopped. For symmetric rotation, there is no lift generated because the flow is perpendicular to the wing, but a large peak in the drag force is observed. In the case of delayed rotation, the incoming flow and the wing position generate a negative lift peak. Adopted from Dickinson et al. (1999).

The computational results of Wang et al. (2004) and Tang et al. (2007) both identify a secondary lift peak after the stroke reversal for the normal-hovering mode (Figure 4.25). However, for the water-treading mode, the results of Tang et al. (2007) show a continuous increase in lift as the airfoil pitch angle increases to its maximum value without a noticeable second peak. The lift generations for both normal and water-treading hovering modes are discussed in more detail in Subsection 4.5.1.

The interpretation of wake-capture force generation has been questioned recently based on the viewpoint that the rotation-independent lift peak is due to a reaction

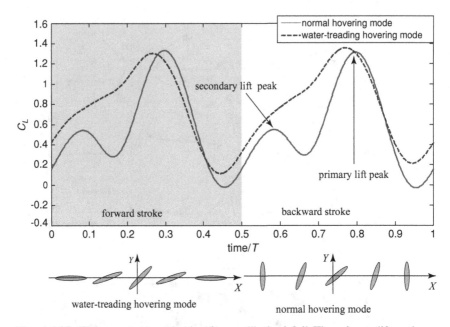

Figure 4.25. Wake-capture mechanism for an elliptic airfoil. The primary lift peaks are mostly generated by dynamic stall, whereas the secondary peaks for normal hovering imply a wake-capture mechanism. For both cases, the Reynolds number is 100, the reduced frequency is 0.357, the stroke amplitude is $1.4c$, and the pitch angle amplitude is 45°.

of accelerating an added mass of fluid (Sunada and Ellington, 2000). In general, the inertia of the flapping wing is increased by the mass of the accelerated fluid, termed added mass (Katz and Plotkin, 2002), which can play a significant role in the aerodynamics of insect flight (Osborne, 1951). The evaluation of the added mass, and thus an estimation of inertial forces, is, however, not easy. Although the mass of a wing itself may be tiny, the mass of the accelerated fluid need not be (Ellington, 1984a; Lehmann, 2004).

4.4.4 Clap-and-Fling Mechanism

One of the most complex kinematic maneuvers in flying animals is the wing–wing interaction of the left and right wings during the dorsal stroke reversal, termed the clap-and-fling mechanism. Weis-Fogh (1973), when studying the flight of the tiny wasp *Encarsia formosa*, found that, at the end of upstroke and at the beginning of the downstroke, the two wings clapped together (clap) and then peeled apart (fling). This mechanism has been further observed by other researchers (Ellington, 1984c; Ennos, 1989; Wootton and Newman, 1979). A modified kinematics termed "clap-and-peel" was found in tethered flying *Drosophila* (Gotz, 1987) and larger insects such as butterflies (Brodsky, 1994), bush crickets, mantises (Brackenbury, 1990), and locusts (Cooter and Baker, 1977). It seems that the clap-and-fling is

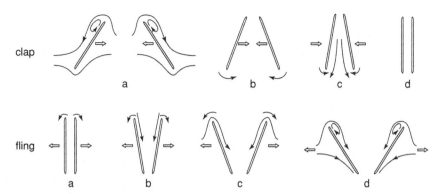

Figure 4.26. Illustration of clap-and-fling mechanism. Adopted from Weis-Fogh (1973).

not used continuously during flight, and more often is observed in insects while carrying loads during a maximum flying performance (Marden, 1987) or performing power-demanding flight turns (Cooter and Baker, 1977). Marden's experiments on various insect species reported that insects with the clap-and-fling wing beat produce about 25% more lift per unit flight muscle (79.2 N kg^{-1} mean value) than insects using conventional wing kinematics (such as flies, bugs, mantids, dragonflies, bees, wasps, beetles, sphinx moths; 59.4 N kg^{-1} mean value).

The clap-and-fling is a close apposition of two wings at the dorsal stroke reversal preceding pronation that is thought to strengthen the circulation during the down-stroke and hence to generate a considerably large lift on the wings. The fling phase preceding the downstroke is thought to enhance circulation that is due to fluid inhalation in the cleft formed by the moving wings, which cause a strong vortex generation at the leading edge. A schematic, shown in Figure 4.26, demonstrates this mechanism. Lighthill (Lighthill, 1973) has shown that a circulation proportional to the angular velocity of the fling was generated. Maxworthy (1979), by a flow-visualization experiment on a pair of wings, reported that, during the fling process, an LEV is generated on each wing and its circulation is substantially larger than that calculated by Lighthill (1973).

Lehmann et al. (2005) used a dynamically scaled mechanical model of the fruit fly, *Drosophila melanogaster*, to investigate force enhancement that is due to contralateral wing interactions during stroke reversal (clap-and-fling). Their results suggest that lift enhancement during clap-and-fling requires an angular separation between the two wings of no more than 10°–12°. Within the limitations of the robotic apparatus, the clap-and-fling augmented total lift production by up to 17%, but the actual performance depended strongly on stroke kinematics. They measured two transient peaks of both lift and drag enhancement during the fling phase: a prominent peak during the initial phase of the fling motion, which accounts for most of the benefit in lift production, and a smaller peak of force enhancement at the end fling when the wings started to move apart. Their investigation indicates that the effect of

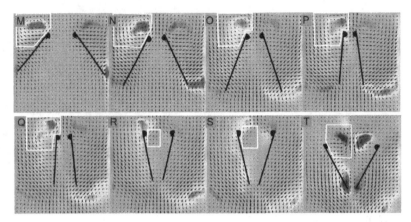

Figure 4.27. Experiment of clap-and-fling by two wings (M–T) using clap-and-fling wing-beat pattern in the robotic wing. Vorticity is plotted according to the pseudo-color code and arrows indicate the magnitude of fluid velocity, with the longer arrows signifying larger velocities, from Lehmann et al. (2005) with permission. (See Plate XX.)

clap-and-fling is not restricted to the dorsal part of the stroke cycle but extends to the beginning of upstroke, suggesting that the presence of the image wing distorts the gross wake structure throughout the stroke cycle (Figure 4.27).

4.4.5 *Wing Structural Flexibility*

As discussed earlier in Chapter 3, wing flexibility improves the fixed-wing MAV performance at a high AoA and allows a better adaptation to the unsteady flight environment by use of passive camber control. The study of a flexible flapping wing is rather complicated because of the kinematic variables, in addition to geometrical and flow variables. Heathcote et al. (2004) experimentally investigated the flexibility effect on thrust generation by a plunging airfoil in hovering conditions (zero free-stream velocity). The airfoil used has a rigid part manufactured from solid aluminum; the flexible part is a steel plate with uniform thickness (see Figure 4.28(c)). The stiffness of the airfoil is controlled by varying plate thicknesses, including 0.05 mm (designated as "very flexible"), 0.1 mm ("flexible"), and 0.4 mm ("rigid"). The airfoil plunging motion follows a sinusoidal law with an amplitude of h_a normalized by the chord. The Reynolds number is defined as $Re = fc^2/\nu$ and varies between 7×10^3 and 2.5×10^4. Here, f is the plunging frequency, c is the airfoil chord, and ν is the fluid kinematic viscosity. They observed that the wing flexibility has an important effect on the thrust generation. For a given plunging amplitude, the very flexible airfoil generates the greatest thrust at low Reynolds numbers and then decreases rapidly as the Reynolds number increases. For Reynolds numbers larger than 10^4, the flexible airfoil generated the largest thrust [Figure 4.28 (a)]. The effect of plunging amplitude on thrust generation for a fixed Reynolds number is illustrated in Figure 4.28(b). Again, the flexible airfoil generated the largest

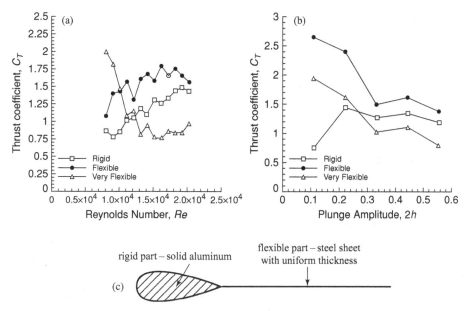

Figure 4.28. Wing-flexibility effect on thrust generation for an airfoil plunging in zero free-stream velocity: (a) thrust coefficient as a function of the Reynolds number for a nondimensional plunging amplitude $h_a/c = 0.194$; (b) variation of thrust coefficient with the plunging amplitude for a Reynolds number of 1.62×10^4; (c) airfoil cross section. Adopted from Heathcote et al. (2004).

thrust, whereas the thrust production of the very flexible airfoil decreased with the plunging amplitude increase. Heathcote et al. (2004) have not found any regime in which the rigid airfoil performed best and suggested that there is an optimum airfoil stiffness that maximizes the thrust for a given plunge amplitude and frequency.

The numerical investigation of a flexible flapping wing needs to solve coupled Navier–Stokes fluid and structural dynamics models with moving boundaries. So far, most of the numerical work adopts simple aerodynamics models. For example, Singh and Chopra (2006) performed numerical aeroelastic analysis for hover-capable, biomimetic flapping wings. The wing dynamics is described with a finite-element-based structural solver, and the unsteady aerodynamic load is based on the assumption that the aerodynamic forces acting on a flapping–pitching wing can be broken down into a number of segments that are accounted for separately and then added to obtain the total force.

Barut et al. (2006) studied the structural behavior of a dragonfly wing, *Aeschna juncea*, under controlled rigid-body motion. Their analysis invokes the corotational form of the updated Lagrangian formulation and utilized the flat triangular shell element. The aerodynamic force is assumed to be known *a priori*. The sequence of motion and deformation with respect to the inertial frame at various time steps is shown in Figure 4.29, and the deformed wing configurations with respect to the body-fixed coordinate are elucidated in Figure 4.30.

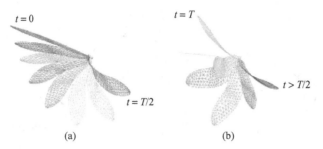

Figure 4.29. Typical deformed configurations of a highly flexible wing as observed from the inertial frame during (a) the first downstroke and (b) upstroke. Adopted from Barut et al. (2006).

Ho et al. (2003) reviewed flapping flight and the application of flow-control technologies. They also developed a coupling method, combining fluid dynamics with the structural dynamics models, and analyzed the aeroelasticity of flapping wings with an attempt to optimize the stiffness distribution for maximum lift and thrust. In their study, two types of wings were tested. One wing had a rigid leading edge, and the other, made of a titanium alloy, had a flexible leading edge. As shown in Figure 4.31, the spanwise stiffness along the leading edge plays an important part in lift production for flapping flight. With wings of the same size, a rigid leading edge produces larger lift coefficients compared with those in which flexible leading edges are used.

From the results of Ho et al. (2003), it seems clear that the stiffness distribution plays an important role in thrust production. They tested two wings with identical configurations. One wing had a paper membrane and the other a Mylar membrane. The paper-membrane wing, which is less flexible than the Mylar wing, produced significantly less thrust than the more flexible Mylar wing (Figure 4.32).

Tests based on different wing designs demonstrate that a stiffer membrane wing does not produce thrust, whereas a more flexible membrane wing does. They suggest that a wing with a rigid outboard frame and a flexible inboard material are desirable for producing both lift and thrust.

The preceding discussion is based on the consideration of isotropic materials, i.e., the materials that exhibit uniform properties along all directions. In reality, biological

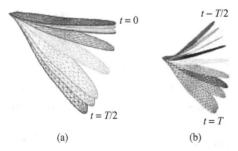

Figure 4.30. Typical deformed configurations of the wing as observed from the body-fixed frame: (a) downstroke, (b) upstroke. Adopted from Barut et al. (2006).

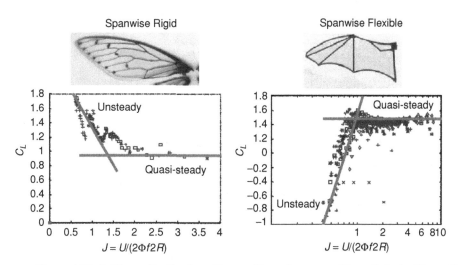

Figure 4.31. Stiffness distribution effect on lift performance. Here C_L is the lift coefficient, J is the advance ratio, U is the forward-flight velocity, Φ is the stroke amplitude, f is the flapping frequency, and R is the wing length. Adopted from Ho et al. (2003).

wings are anisotropic in their mechanical properties. Combes and Daniel (2003) investigated wing flexibility by measuring the flexural stiffness EI (where E is Young's modulus and I is the moment of inertia) of wings in both the spanwise and chordwise directions in 16 insect species from six orders. The forewings from insects used in the study are shown in Figure 4.33. Flexural stiffness is a composite measure of the overall bending stiffness of a wing; it is the product of the material stiffness E and the second moment of the wing I. Their measurements have shown that the spanwise stiffness scales with the cube of the wingspan, whereas the chordwise flexural stiffness

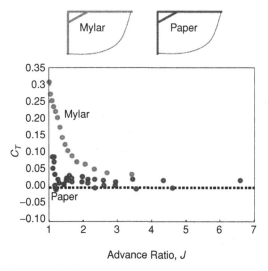

Figure 4.32. Stiffness effect on thrust production. Here C_T is the thrust coefficient and J is the advance ratio as defined in Figure 4.31. Adopted from Ho et al. (2003).

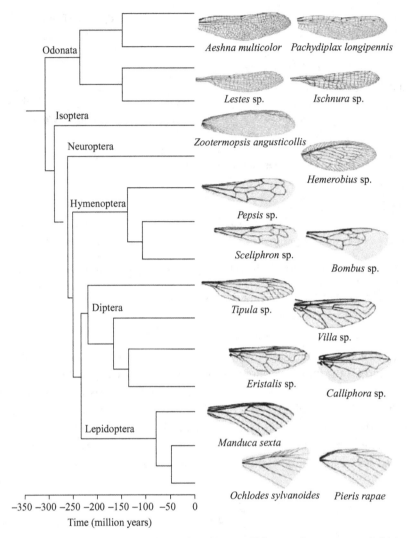

Figure 4.33. Wing structures for various insects. Veins are drawn at actual thicknesses; wings are not shown to scale. Genus and species are shown under each wing. Adopted from Combes and Daniel (2003).

scales with the square of the chord length (see Figure 4.34). This anisotropy is due to a common feature of insect wings: leading-edge veins.

In a recent effort, Raney and Slominski (2004) investigated mechanization and control concepts with applications to a resonant flapping MAV, as many natural flyers generate lift by using resonant excitation of their aeroelastic tailored structures. Their structural dynamic model of flapping-wing structures for MAVs requires the judicious combination of several structural elements. The flapping-wing structure resembling the wing of a hummingbird, developed by Raney and Slominski, is shown in Figure 4.35 and consists of the following structural elements: (i) composite beams undergoing moderate deflection; (ii) composite plates undergoing deflection; (iii) anisotropic flexible membranes undergoing large deformation; and (iv) wire-type

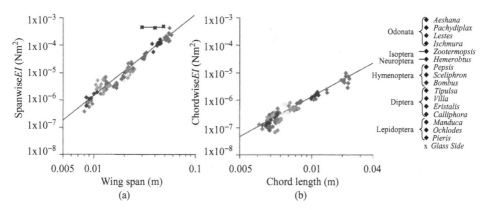

Figure 4.34. Flexural stiffness versus span/chord length in 16 insect species. Axes are on a logarithmic scale: (a) spanwise flexural stiffness EI versus wingspan; (b) chordwise flexural stiffness EI versus chord length. Adopted from Combes and Daniel (2003).

elements that connect the end of the beam at the trailing edge to preserve the shape of the wing. In addition, to study a means of changing the flight modes and generating maneuvers, Raney and Slominski investigated models that include the inertia loads coming from the rigid-body dynamics of the MAV as well as the kinematic input provided at the root of the flapping wing at the hinge point. They used a vibratory system that follows the basic arrangement of the skeletal and muscular systems that drive a typical bird wing. By supplying appropriate amplitude, phasing, and time-varying

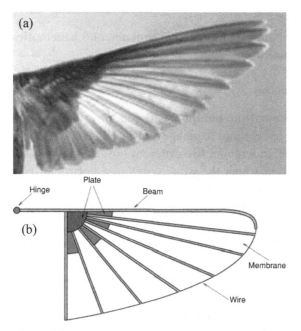

Figure 4.35. Biologically inspired structure for flexible flapping wings: (a) extended hummingbird wing; (b) schematic of the artificial wing structure including beam, plates, membrane, and wire elements. Adopted from Raney and Slominski (2004).

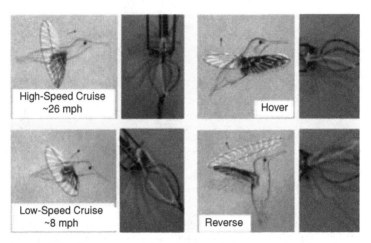

Figure 4.36. Comparison of the wingtip trajectories produced by the vibratory flapping system with those exhibited by hummingbirds in various flight modes. Adopted from Raney and Slominski (2004). (See Plate XXI.)

forces similar to the inputs of the bird muscles, they obtained wingtip trajectories similar to those observed in hummingbirds (see Figure 4.36). They observed that the transitions between wing-beat patterns are fast, being accomplished in about four flapping cycles. These results show that flight-mode changes required by an agile flapping-wing MAV can be made by a biologically inspired mechanism that provides sufficient control over the vibratory wingtip trajectories.

4.5   Effects of Reynolds Number, Reduced Frequency, and Kinematics on Hovering Aerodynamics

4.5.1 *Hovering Kinematics*

In this section, we present the interplay between hovering kinematics and fluid physics to gain further insight into flapping-wing aerodynamics. Specifically, a 15% thickness elliptic airfoil undergoing two different hovering modes is studied. The normal hovering mode, in which the wing moves in a horizontal plane, is a mode popularly used by insects and small birds in hovering. Wang et al. (2004) and Tang et al. (2007) used single-harmonic kinematics for both plunging and pitching motions. The airfoil rotation is symmetric, i.e., the center of rotation is the center of the elliptic airfoil. The water-treading mode (Freymuth, 1990) is also considered. For both hovering modes, the flapping motion and the rotational motion are described by Eqs. (4.13) and (4.14), and a schematic of the airfoil movement is presented in Figures 4.3(a) and 4.3(b).

First, experimental measurements and numerical predictions of the normal hovering mode results for an initial AoA, $\alpha_0 = 90°$, a pitch angle amplitude, $\alpha_a = 45°$, a nondimensional stroke amplitude, $h_a/c = 1.4$, a phase lag, $\varphi = 90°$, a reduced

Table 4.2. *Kinematic parameters for water-treading and normal hovering modes (the Reynolds number for both cases is 100)*

Hovering mode	Initial AoA α_0	Pitch amplitude α_a	Stroke amplitude h_a/c	Reduced frequency k	Phase difference
Water treading	$0°$	$45°$	1.4	1/2.8	$-\pi/2$
Normal	$90°$	450	1.4	1/28	$\pi/2$

frequency, $k = 0.357$ (as defined in Eq. (4.12)), and a Reynolds number of 75 are compared with similar computational and experimental results. Figure 4.37 shows the present computational results and those of Wang et al., together with experimental results of Birch and Dickinson (Wang et al., 2004). The current results show good overall agreement between the experiments and the computations.

Next, we compare the aerodynamic forces generated by the airfoil undergoing the two hovering modes. To make the comparison possible, consistent kinematic parameters are selected, as presented in Table 4.2. As defined earlier (Section 4.1), the Reynolds number is based on the maximum plunging velocity and the airfoil's chord and is calculated to be 100.

Figure 4.38 shows the lift and drag coefficients during one complete cycle for the water-treading and normal hovering modes. To illustrate the unsteady effects, the quasi-steady value of the normal hovering mode (according to Eq. (16) in Wang et al. (2004)) is also included. Clearly, even if one introduces correction factors to adjust the quantitative values, the existence of multiple aerodynamic peaks and their phase angles cannot be captured by the quasi-steady model.

In the case of the water-treading hovering mode, for the first half of the forward stroke, the airfoil accelerates and pitches-up. During this interval, the lift increases constantly (Figure 4.38(a), t1–t3), and the unsteady dynamics results in delayed flow separation even at instantaneously high AoAs, as indicated by the vorticity contours

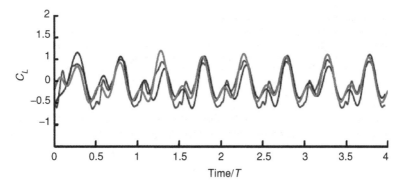

Figure 4.37. Numerical and experimental results of the flapping motion of a fruit fly: red, experimental results of Dickinson and Birch (Wang et al., 2004); Blue, numerical solution of Wang et al. (2004); green, numerical solution of Tang et al. (2007). $h_a/c = 1.4$, $\alpha_a = 45°$, $Re_{f2} = 75$, $k = 0.357$. (See Plate XXII.)

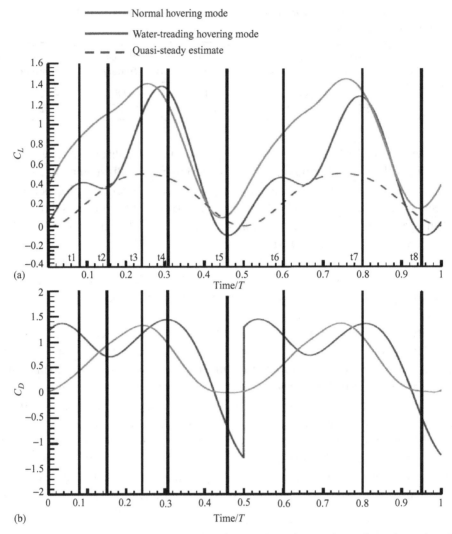

Figure 4.38. One cycle force history for two hovering modes and quasi-steady value of normal hovering mode. $h_a/c = 1.4$, $\alpha_a = 45°$, $k = 0.357$, and $Re_{f2} = 100$. (a) Lift coefficient, (b) drag coefficient. The selected normalized time instants are t1 = 0.08, t2 = 0.17, t3 = 0.25, t4 = 0.31, t5 = 0.45, t6 = 0.60, t7 = 0.80, t8 = 0.94. (See Plate XXIII.)

plotted in Figure 4.39, t1–t3. The maximum lift is reached close to the middle of the half-stroke around the instant when the pitch angle reaches the highest value (Figure 4.39, t3). However, as indicated in Figure 4.38, their correspondence is not exact. This confirms the well-known point that flapping aerodynamics cannot be correctly accounted for by steady-state aerodynamics theory. Beyond this stage, the airfoil starts to decelerate and pitches-down. The flow separates and a large recirculation bubble forms on the upper side of the airfoil (Figure 4.39, t4 and t5), leading to a decrease in lift to the minimum value (Figure 4.38(a), at time t5). The same pattern is repeated for the backward stroke.

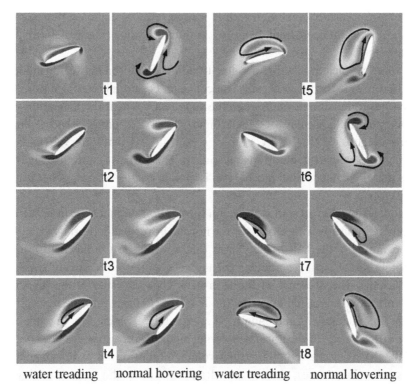

water treading   normal hovering   water treading   normal hovering

Figure 4.39. Vorticity contours for two hovering modes. $h_a/c = 1.4$, $\alpha_a = 45°$, $k = 0.357$, and $Re_{f2}=100$. Red, counterclockwise vortices; blue, clockwise vortices. The flow snapshots (t1 to t8) correspond to the time instants defined in Figure 4.38. Adopted from Tang et al. (2007). (See Plate XXIV.)

For the normal hovering mode, at the beginning of the forward stroke, the airfoil accelerates and pitches-down. The rotation of the airfoil speeds up the flow around the leading and trailing edges, creating a suction zone on the upper side of the airfoil, and the high-pressure stagnation area on the lower side is increased because of the fluid driven from the surroundings by the previously formed vortex (Figure 4.39, t1). This combination of low- and high-pressure areas leads to an increase in lift at the beginning of the stroke (Figure 4.38(a), at time t1). As the airfoil rotates downward more and accelerates, the fluid is accelerated toward the trailing edge and the high-pressure stagnation area decreases (Figure 4.39, t2) and so does the lift, reaching a local minimum at time/$T \sim 0.17$ for the forward stroke and 0.57 for the backward stroke, as shown in Figure 4.38(a). Around the middle of each half-stroke, the airfoil travels at almost constant pitch angle. A recirculation bubble attached to the airfoil forms on the upper surface (Figure 4.39, t3, t4, t5, around time/$T \sim 0.3$ and 0.8) and helps increase the lift and drag to their maximum values during one complete stroke (Figures 4.38(a) and 4.38(b), at t4 and t7). After the maximum pitch angle and translation velocity are reached (time/$T = 0.25$ and 0.75) during one half-stroke, the airfoil decelerates and pitches-up, leading to flow separation on the upper side of the airfoil (Figure 4.39, t5 and t8). The detachment of the large vortical structure from the upper airfoil surface combined with rapid deceleration decreases the circulation, and therefore the lift coefficient drops to its minimum value (Figure 4.38(a), at times t5 and t8).

The force coefficient history for water-treading and normal hovering modes indicates differences in the lift-generation mechanism. For both hovering modes, the lift force reaches its maximum value when the airfoil moves near the maximum velocity and moderate AoA. Similar lift peaks (Figure 4.38(a), at times t4 and t7) and flow structures (Figure 4.39, t4 and t7) are observed in this time interval (midstroke), suggesting the same lift-generation mechanism. The vorticity contours (Figure 4.39) indicate that the delayed-stall mechanism is mainly responsible for generating most of the lift force.

Although the delayed stall is the main lift-generation mechanism in the case of the water-treading hovering mode, for the normal hovering mode, the local lift peaks at the beginning of the half-strokes show that a wake-capture mechanism is also a contributing factor (it gives about a 0.2 increase to the quasi-steady value of the lift coefficient). The presence of the twin-peak characteristics of the lift and drag time histories in the normal hovering mode again confirms that the fluid physics is distinctly time dependent and cannot be adequately explained by the steady-state theory. Furthermore, for the normal hovering mode, the drag pattern does not mimic that of the lift, as evidenced by the relative magnitudes of the two peaks in lift and drag histories. In contrast, the lift and drag patterns in the water-treading mode show much stronger correspondence, further suggesting the role played by the wake-capture mechanism in the normal hovering mode. Hence, depending on the detailed kinematics, the lift-generation mechanisms at $Re_{f2} = 100$ exhibit different physical mechanisms.

The average lift coefficient for both cases is computed as the summation of the lift coefficient over the last three periods divided by the total time. For the water-treading hovering mode, an average lift coefficient of 0.77 is obtained; for the normal hovering mode, the average lift coefficient is 0.56, suggesting that water-treading mode performs better at $Re_{f2} = 100$ for the given kinematics parameters. However, other aspects, such as the Reynolds number, detailed geometry as well as the kinematic parameters, need to be considered before comprehensive comparsions of the alternative flapping dynamics can be conducted.

4.5.2 *Scaling Effect on Force Generation for Hovering Airfoils*

As discussed in Chapter 1 and the first section of this chapter, the natural flyers operate in regimes where both inertial and viscous forces are important. In the following subsections, we present selected case studies to highlight the scaling effect on aerodynamic force-production mechanisms.

2D Water-Treading Hovering Mode   To investigate the Reynolds number effect on aerodynamic forces and the flow structure, we computed the hovering aerodynamics of the water-treading mode at $Re_{f2} = 100$ and $Re_{f2} = 1700$. Based on

Table 4.3. *Parameters for the water-treading hovering mode used for Reynolds number effect study*

Hovering mode	Initial AoA α_0	Pitch angle amplitude α_a	Stroke amplitude h_a/c	Reduced frequency k	Phase difference φ	Reynolds number Re_{f2}
Water treading	0°	45°	1.4	1/2.8	−π/2	100
Water treading	0°	45°	1.4	1/2.8	−π/2	1700

the same kinematics of the $Re_{f2} = 100$ case, the aerodynamics of the water-treading mode is assessed. The kinematics and flow parameters for these cases are summarized in Table 4.3, and the airfoil motion schematic is presented in Figure 4.3.

The lift coefficients for water-treading hovering mode at Reynolds numbers of 100 and 1700 are plotted in Figure 4.40. The lift coefficient peaks are noticeably higher for the Reynolds number of 1700 than that for the Reynolds number of 100. Although the force patterns between the two Reynolds numbers are similar, the higher Reynolds number case exhibits larger differences in the lift peak values between forward and backward strokes.

The pressure distributions on the airfoil surface, plotted in Figure 4.41, show that, near the maximum lift peaks, the high-pressure stagnation area on the lower side of the airfoil is similar in both shape and magnitude for the two Reynolds number studied. However, on the upper side of the airfoil, the mild variation of the pressure gradient for the low Reynolds number case [Figure 4.41(a), times t1 and t3] suggests that the flow is attached, whereas for the high Reynolds number [Figure 4.41(b), times t1 and t3] the low-pressure area near the leading edge indicates a recirculation zone corresponding to the LEV [Figure 4.42(b), and 4.42(d) at times t1 and t3]. This

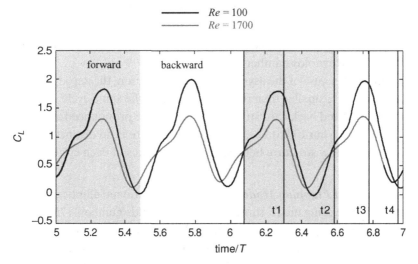

Figure 4.40. Lift coefficient for the water-treading mode. $h_a/c = 1.4$, $\alpha_a = 45°$, $k = 0.357$, and Reynolds numbers of 100 and 1700. The selected normalized time instants are t1 = 6.25, t2 = 6.48, t3 = 6.77, t4 = 6.97. (See Plate XXV.)

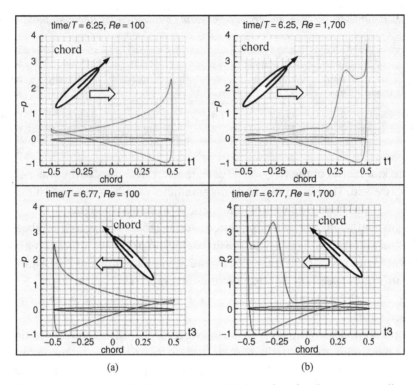

Figure 4.41. Pressure distribution on the airfoil surface for the water-treading mode. $h_a/c = 1.4$, $\alpha_a = 45°$, $k = 1/2.8$. (a) Reynolds number = 100, (b) Reynolds number = 1700. The flow figures (t1, t3) correspond to the time instants defined in Figure 4.40.

low-pressure area is responsible for most of the high lift peak values seen in the case of a Reynolds number of 1700.

At a Reynolds number of 1700, because of the smaller dissipation rate, the vortices sustain the effects of the asymmetric starting condition (the wing motion starts moving from one end). This phenomenon indicates the important role of viscous stress in the low Reynolds number regime.

In summary, because of the asymmetric start condition, the aerodynamic force in one stroke is a little smaller than the other stroke in the same cycle. The difference between forward and backward strokes becomes more pronounced as the Reynolds number increases from 100 to 1700. Nevertheless, there is no distinctive, qualitative difference in the flow structure between the two strokes of each cycle.

2D Normal Flapping Mode For the aforementioned elliptical airfoil following the normal hovering mode, three different Reynolds numbers (75, 300, and 500) were studied by Tang et al. (2007). In the following discussion, the motion parameters are same as the cases in Table 4.2 except that the flapping amplitude h_a and the reduced frequency k are changed to match the designated Reynolds number. In Figure 4.43, the lift coefficients at the three Reynolds numbers are shown. It is clear that the aerodynamic forces during the forward and backward strokes are symmetric

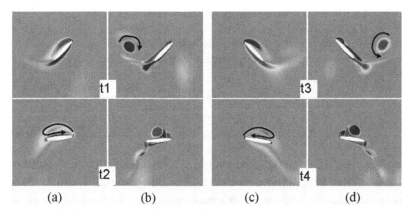

Figure 4.42. Vorticity contours for the water-treading mode. $h_a/c = 1.4$, $\alpha_a = 45°$, $k = 0.357$, Red, counterclockwise vortices; blue, clockwise vortices. (a), (c) Reynolds number $= 100$; (b), (d) Reynolds number $= 1700$. The flow figures (t1–t4) correspond to the time instants defined in Figure 4.40. (See Plate XXVI.)

at $Re_{f2} = 75$; at $Re_{f2} = 300$ and 500 the lift coefficient variations become asymmetric between the forward and backward strokes of each cycle. It should be emphasized that the aerodynamic characteristics regarding the Reynolds number effect are highly dependent on the kinematic parameters. For example, the qualitative force patterns in the normal hovering mode are quite different from those in the water-treading mode. For the water-treading hovering mode, although quantitative differences can be observed as the Reynolds number increases from 100 to 1700, qualitatively, as shown in Figure 4.40, similar force patterns are observed. For the normal hovering mode, the aerodynamic force patterns between $Re_{f2} = 75$ and 500 are qualitatively different, suggesting that different physical mechanisms exist. In Figure 4.44, the flow structures of the corresponding positions between the forward and backward strokes at $Re_{f2} = 300$ are plotted. The vortex pair below the airfoil in Figure 4.44(a) is not found in Figure 4.44(c) (corresponding to the backward stroke at the same position and AoA). Figure 4.45 shows that, under the normal mode, there is only one pair of vortices around the airfoil at $Re_{f2} = 75$ whereas there are two pairs of vortices interacting at $Re_{f2} = 500$. To quantify this asymmetric phenomenon caused by the history effect, the difference of the average lift and drag coefficients of the two forward and backward strokes in each cycle, for both normal and water-treading modes, are listed in Table 4.4.

Table 4.4 suggests that, for the normal hovering mode, the differences in lift and drag between forward and backward strokes increase with the Reynolds number.

4.6 Aerodynamics of a Hovering Hawkmoth

The 3D numerical simulations of flapping wings have become a common tool in investigating flapping-wing aerodynamics. Isogai et al. (2004) studied the hovering flight of the dragonfly, *Anax parthenope julius*, by using a 3D Navier–Stokes solver. They found that varying the phase angle between the flapping of the forewing and

Table 4.4. *Differences in average lift and drag coefficients between forward and backward strokes for the normal hovering mode at different Reynolds numbers with a flapping amplitude of $h_a/c = 0.25$*

Aerodynamic coefficient	$Re_{f2} = 75$	$Re_{f2} = 300$	$Re_{f2} = 500$
ΔC_L	0.002	0.325	0.330
ΔC_D	0.045	0.105	0.125

hindwing has little effect on the time-averaged force production. Liu and coworkers (Aono et al., 2006; Liu, 2005) developed a biology-inspired dynamic flight simulator by using realistic body–wing morphologies, flapping-wing kinematics, and a fluid dynamics model. In the following subsections, we present results to highlight the interaction of the vortical structures and the unsteady flow field for a hovering hawkmoth.

4.6.1 *Downstroke*

Figure 4.46 shows isovorticity surfaces above the flapping wings and body over one flapping cycle. During the early downstroke stage, a vortex structure around the wing edge is created, accompanying the flapping-wing motion. When the wings

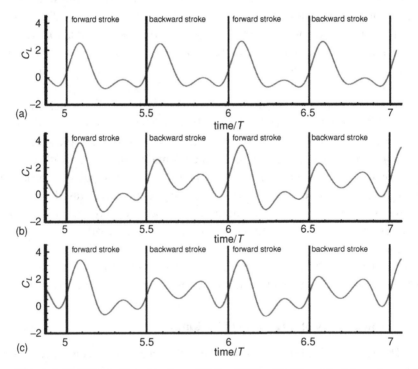

Figure 4.43. Lift coefficients of an elliptic airfoil with the normal hovering mode at different Reynolds numbers: (a) $Re_{f2} = 75$; (b) $Re_{f2} = 300$; (c) $Re_{f2} = 500$. In all cases, the stroke amplitude is $h_a/c = 0.25$.

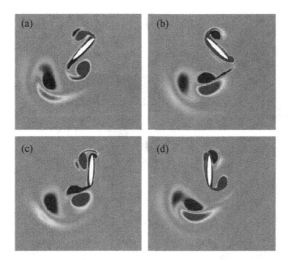

Figure 4.44. Vorticity contours at two corresponding positions during (a), (c) forward and (b), (d) backward strokes. Stroke amplitude $h_a/c = 0.25$, pitch angle amplitude $\alpha_a = 45°$, and $Re_{f2} = 300$. (See Plate XXVII.)

accelerate, the vortex evolves into three entities, namely, the LEV, the trailing-edge vortex (TEV), and the wingtip vortex (WTV). As illustrated in Figure 4.46(a), the LEV with the maximum size on the upper surface of a wing is detected when the positional angle of the wing approaches zero in the middle of the downstroke. The LEV is broken down at the location 70%–80% of the wing length distal from the wing base. The TEV can detach from the wing along with the acceleration of the wing during the downstroke, and then it connects to the WTV. The LEV, the WTV, and the TEV connect with each other to form a continual vortex chain along the wing edge. Such a vortex chain can facilitate the downward flow and thereby improve the effectiveness of aerodynamic force generation.

4.6.2 *Supination*

When the positional angle of the wings changes from $-36.7°$ to $-46.4°$ while the AoA of the wings changes from $-17.2°$ to $60.2°$ [Figure 4.46(b)], the preceding

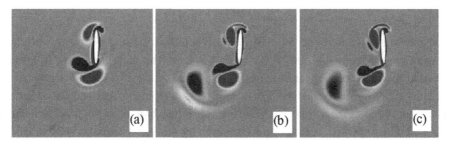

Figure 4.45. Vorticity contours at time $/T = 5.5$ and three different Reynolds number with a stroke amplitude $h_a/c = 0.25$ and $\alpha_a = 45°$: (a) $Re_{f2} = 75$; (b) $Re_{f2} = 300$; (c) $Re_{f2} = 500$. (See Plate XXVIII.)

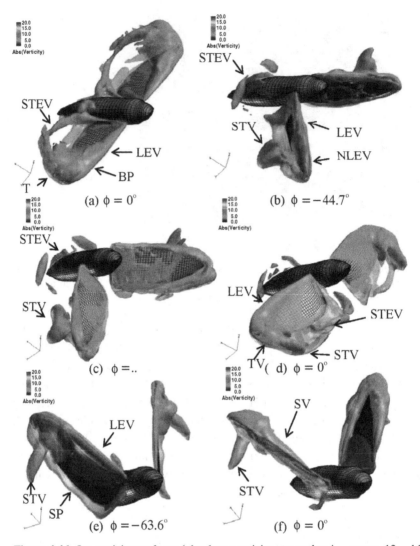

Figure 4.46. Isovorticity surfaces (absolute vorticity strengths: 4 = green, 13 = blue) around flapping wings and body of a hawkmoth during a flapping cycle. Shedding WTV, STV; shedding TEV, STEV; new LEV, NLEV; stopping vortex, SPV; starting vortex SV; and breakdown point, BP. (See Plate XXIX.)

LEV still stays attached and a new LEV is created, stretching from the wingtip to the wing base. Meanwhile, the WTV detaches from the wing and stays in an area immediately distal to the back side of the wing. The shedding TEV flows toward the body hip along the body surface. Furthermore, a stopping vortex is created around the trailing edge (Figure 4.46(b)). During the latter half of supination, when the wing translates upward, rotates, and increases its AoA, a vortex is created around the wing edge (Figure 4.46(c)), from the wingtip to the wing base. Furthermore, the wake capture is observed, which, at supination, seems less influential on the aerodynamic force generation (Figure 4.47).

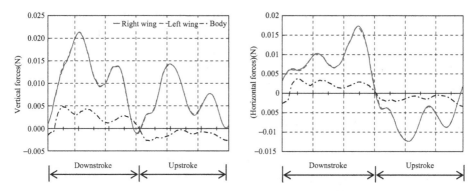

Figure 4.47. Time variation of vertical and horizontal forces during a flapping cycle of a hovering hawkmoth. Right wing, green solid curve; left wing, red broken curve; and body, black dashed-dotted curve.

4.6.3 *Upstroke*

As illustrated in Figure 4.46 (d), the LEV with the maximum size appears again when the positional angle of the wing is nearly zero during the upstroke. Although the appearance of a large-scale LEV is similar to that during the downstroke, the size of the LEV is evidently smaller here. Because the size of the LEV directly influences lift generation, it is expected that the upstroke is less effective than the downstroke in lift force generation. The large-scale WTV and the shedding of the TEV accompanying the wing acceleration are also predicted. Moreover, the breakdown of the LEV does not occur because of the difference in WTV shape and LEV size.

4.6.4 *Pronation*

As the positional angle of the wing ranges from 51.6° to 48.7° whereas the AoA of the wing varies between 46.2° and −46.2°, the LEV and the WTV interact with each other around 70%–80% of the wing length distal from the wing base. This makes the LEV unstable, and subsequently the LEV begins to shed off. The continuous WTV on the wing is observed up to the early pronation. In addition, a stopping vortex is found around the wing edge (Figure 4.46(e)). In the later phase of the pronation, a starting vortex is created around the wing edge and the WTV is shed out and pushed toward the body hip. The presence of wake capture is not predicted at the pronation (Figure 4.46(f)) and hence the aerodynamic forces are estimated to be insignificant (Figure 4.47).

4.6.5 *Evaluation of Aerodynamic Forces*

Low-pressure regions on the wing surface are detected beneath the vortices, which contribute substantially to lift generation. As discussed, the attached LEV can enhance lift during downstroke and upstroke. The instantaneous vertical force (lift) and horizontal force (drag) are estimated and plotted in Figure 4.47. The computed results show that a substantial portion of the lift of a hovering hawkmoth is produced during the translational phase of wing motion, i.e., downstroke and upstroke. Specifically, at both the mid-downstroke and the mid-upstroke the instantaneous vertical force reaches the maximum value because of the large-scale LEVs (Figures 4.46(a) and 4.46(d)). The reason that the peak vertical force generated during the upstroke is lower than that generated during the downstroke is because a more intense LEV is produced during the downstroke (Figures 4.46(a) and 4.46(d)).

The computed results demonstrate that the aerodynamic force production of the hovering hawkmoth follows a general pattern in that the lift is produced largely during the downstroke and partially during upstroke with delayed stall, and the drag is produced also largely during downstroke. In contrast, the thrust is produced largely during upstroke. This instantaneous force-generation pattern is similar to that of a hummingbird (Warrick et al., 2005). Moreover at the midpronation and midsupination, the aerodynamic forces are much smaller because there is no attached LEV on the wing surface (Figures 4.46(b), 4.46(c), 4.46(e) and 4.46(f)). Based on the 3D flow simulations, the average lift force is estimated to be 17 mN over a flapping cycle, which is comparable to the weight (14.7 mN) of a hawkmoth. Further research is needed to further ascertain lift and thrust generation during a flapping cyle, as functions of sizing, namely, the Reynolds number and reduced frequency (or, in forward flight, the Strouhal number).

4.6.6 *Aerodynamic and Inertial Powers of Flapping Wings*

Based on the computed instantaneous aerodynamic forces and the wing velocities, the instantaneous inertial and aerodynamic powers are calculated over a cycle of the flapping wings (Figure 4.48). The aerodynamic power P_{aero} reaches a maximum in the late phase of both the downstroke and the upstroke. During the translational phase of the wing motion, the insect generates substantial aerodynamic power because of high aerodynamic force production. The mean aerodynamic power is estimated to be 90 W/kg, which is approximately the same as that of the experimental results.

The inertial power P_{iner} is the power needed to accelerate the mass of the wing. It is lower in magnitude compared with the aerodynamic power and shows a different time-variation pattern. P_{iner} increases as the wing accelerates and decreases as it decelerates. Note that the negative sign of P_{iner} means that the direction of the forces acting on the wing is opposite to that of the wing velocity. P_{iner} exhibits a negative value in the main part of the downstroke and the upstroke except for the wing-acceleration phase. The computed mean inertial power required for accelerating the wing amounts to 65 W/kg. Note that we assume that the wing deceleration accrues with no cost and that there is no elastic storage.

Total muscle-mass-specific mechanical power P_{tot} is the mechanical power required for, moving the wings. It is calculated by the summation of P_{iner} and P_{aero}. Peaks of P_{tot} appear in the early phase of the downstroke and the upstroke. P_{tot} becomes negative in the late upstroke as the decelerating wings produce less power

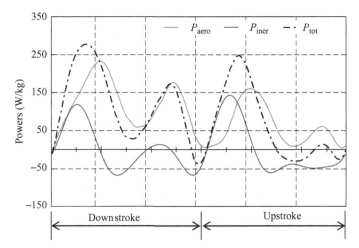

Figure 4.48. Time variation of the muscle-mass-specific flight powers during a flapping cycle of a hovering hawkmoth. P_{aero}, orange; P_{iner}, green; P_{tot}, black dashed-dotted curve.

than is required for overcoming the inertial forces. The computed time course of P_{tot} is in qualitative agreement with the experimental results (Liu et al., 1998).

4.7 Concluding Remarks

In this chapter we described prominent issues on biological flapping flight. Although simulation-based biological fluid dynamics has been fast in developing, the scope covered is still quite limited. Similarly, there is substantial room for further experimental investigations.

1. Various lift-enhancement mechanisms were discussed, including LEV, rapid pitch-up, wake capture, and clap-and-fling.
2. It is known that, depending on the flapping-wing kinematics, stroke amplitude, frequency, Reynolds number, and the free-stream environment, different flow structures result, which can lead to either drag or thrust to be generated. Clearly, flapping kinematics has a strong influence on time-dependent aerodynamics, resulting in a variety of patterns and characteristics. Distinctly different aerodynamic performances are observed between the different kinematic modes at the same Reynolds number and with the same mode but between different Reynolds numbers. To develop a suitable knowledge base and design guidelines for flapping-wing MAVs, a thorough understanding of the kinematics, large vortex structures, and Reynolds numbers is essential as these processes directly influence the lift and thrust generation.
3. Although the flapping and associated fluid flows are intrinsically time dependent, quasi-steady models were used to capture some basic physics for certain flyers. The quasi-steady approach was also used to estimate the mechanical power requirements of hummingbirds and bumblebees. However,

high-fidelity computational aerodynamics models also indicate that history effect and kinematics strongly influence the aerodynamics. Therefore there is a clear need to establish a domain of applicability for the approximate methods and to develop better models that take into account these effects for prediction of performance and power requirements of flapping-wing MAVs.

4. It appears that structural flexibility can delay stall and enhance aerodynamics in fixed wings. Anisotropic wings, with a spanwise bending stiffness about 1 to 2 orders of magnitude larger than the chordwise bending one, are observed in biological flyers. If torsion stiffness can be tailored over the plane of the wing, what wing kinematics can yield optimum thrust generation?

5. Wind gusts create intrinsic unsteadiness in the flight environment. Fundamentally, the characteristic flapping time scale of insects and hummingbird (tens to hundreds of hertz) is much shorter than the time scale of a typical wind gust (around 1 Hz). Hence, from the flapping-wing time scale, many wind gust effects can be treated in a quasi-steady manner. However, the vehicle control system (as in the case of a biological flyer) operates at lower frequencies, and its time scales are comparable with those of anticipated wind gusts. Therefore there is a clear multiscale problem among unsteady aerodynamics, wind gust, and vehicle control dynamics.

6. Not only flapping wing aerodynamics but also the dynamics of inertia can play an important role in animal locomotion.

References

Albertani, R., Hubner, P., Ifju, P. G., Lind, R., and Jackowski, J. (2004). Wind tunnel testing of micro air vehicles at low Reynolds numbers, SAE Paper 2004-01-3090, presented at the *SAE 2004 World Aviation Conference*, Reno, NV.

Alexander, D. E. (2002). *Nature's Flyers* (Baltimore/London, Johns Hopkins University Press).

Alexander, R. M. (1976). Mechanics of bipedal locomotion, in P. S. Davies (Ed.), *Perspectives in Experimental Biology* (Oxford, Pergamon Press), pp. 493–504.

Alexander, R. M. (1997). The U J and L of bird flight, *Nature (London)* **390**, 13.

Anders, J. B. (2000). Biomimetic flow control, *AIAA Paper 2000-2543*.

Anderson, Jr., J. D. (1989). *Introduction to Flight* (New York, McGraw-Hill).

Anderson, J. M., Streitlien, K., Barrett, D. S., and Triantafyllou, M. S. (1998). Oscillating foils of high propulsive efficiency, *Journal of Fluid Mechanics* **360**, 41–72.

Aono, H., Liang, F., and Liu, H. (2006). Near- and far-field aerodynamics in insect hovering flight: An integrated computational study, *Journal of Experimental Biology* (submitted).

Aymar, G. C. (1935). *Bird Flight* (New York, Dodd and Mead).

Azuma, A. (1983). *Local Momentum and Local Circulation Methods for Fixed, Rotary and Beating Wings*, Thesis, Institute of Interdisciplinary Research, Faculty of Engineering (Tokyo, University of Tokyo).

Azuma, A. (1992). *The Biokinetics of Flying and Swimming* (Tokyo, Springer-Verlag).

Barut, A., Das, M., and Madenci, E. (2006). Nonlinear deformations of flapping wings on a micro air vehicle, *AIAA Paper 2006-1662*.

Bass, R. L., Johnson, J. E., and Unruh, J. F. (1982). Correlation of lift and boundary-layer activity on an oscillating lifting surface, *AIAA Journal* **20**, 1051–6.

Bechert, D. W., Bruse, M., Hage, W., and Meyer, R. (1997). Biological surfaces and their technological application–laboratory and flight experiments on drag reduction and separation control, *AIAA Paper 97-1960*.

Berger, M. A. M. (1999). Determining propulsive force in front crawl swimming: A comparison of two methods, *Journal of Sports Sciences* **17**, 95–105.

Betz, A. (1912). Ein Beitrag zur Erklarung des Segelfluges, *Zeitschrift für Flugtechnik und Motorluftschiffahrt* **3**, 269–72.

Biewener, A. A. (2003). *Animal Locomotion*, Oxford Animal Biology Series (Oxford, Oxford University Press).

Birch, J. M. and Dickinson, M. H. (2001). Spanwise flow and the attachment of the leading-edge vortex on insect wings, *Nature (London)* **412**, 729–33.

Birch, J. M., Dickson, W. B., and Dickinson, M. H. (2004). Force production and flow structure of the leading edge vortex on flapping wings at high and low Reynolds numbers, *Journal of Experimental Biology* **207**, 1063–72.

Bohorquez, F., Rankins, F., Baeder, J., and Pines, D. (2003). Hover performance of rotor blades at low Reynolds numbers for rotary wing micro air vehicles. An experimental and CFD study, *AIAA Paper 2003–3930*.

Brackenbury, J. (1990). Wing movements in the bush cricket *Tettigonia viridissima* and the mantis *Ameles spallanziana* during natural leaping, *Journal of Zoology* **220**, 593–602.

Bradley, R. G., Smith, C. W., and Wary, W. O. (1974). An experimental investigation of leading-edge vortex augmentation by blowing, *NASA CR-132415*.

Bratt, J. B. (1953). Flow patterns in the wake of an oscillating airfoil, *Aeronautical Research Council Technical Report R and M 2773*.

Brodsky, A. K. (1994). *The Evolution of Insect Flight* (New York, Oxford University Press).

Brown, W. C. (1939). Boston low-speed wind tunnel, and Wind tunnel: Characteristics of indoor airfoils, *Journal of International Aeromodeling*, 3–7.

Buckholz, R. H. (1986). The functional role of wing corrugation in living system, *Journal of Fluids Engineering* **108**, 93–7.

Campbell, J. F. (1976). Augmentation of vortex lift by spanwise blowing, *Journal of Aircraft* **13**, 727–32.

Carmichael, B. H. (1981). Low Reynolds number airfoil survey, *NASA CR 1165803*.

Cebeci, T. (1988). Essential ingredients of a method for low Reynolds-number airfoils, *AIAA Journal* **27**, 1680–8.

Chai, P. and Dudley, R. (1996). Limits to flight energetics of hummingbirds hovering in hypo-dense and hypoxic gas mixtures, *Journal of Experimental Biology* **199**, 2285–95.

Chai, P. and Millard, D. (1997). Flight and size constraints: Hovering performance of large hummingbirds under maximal loading, *Journal of Experimental Biology* **200**, 2757–63.

Chambers, L. L. G. (1966). A variational formulation of the Thwaites sail equation, *Quarterly Journal of Mechanics and Applied Mathematics* **19**, 221–31.

Chasman, D. and Chakravarthy, S. (2001). Computational and experimental studies of asymmetric pitch/plunge flapping – The secret of biological flyers, *AIAA Paper 2001-0859*.

Chen, K. K. and Thyson, N. A. (1971). Extension of Emmons' spot theory to flows on blunt bodies, *AIAA Journal* **9**, 821–5.

Childress, S. (1981). *Mechanics of Swimming and Flying* (New York, Cambridge University Press).

Cloupeau, M. (1979). Direct measurements of instantaneous lift in desert locust; Comparison with Jensen's experiments on detached wings, *Journal of Experimental Biology* **80**, 1–15.

Collins, P. Q. and Graham, J. M. R. (1994). Human flapping – Wing flight under reduced gravity, *Aeronautical Journal* **98**, 177–84.

Combes, S. A. and Daniel, T. L. (2003). Into thin air: Contributions of aerodynamic and inertial-elastic forces to wing bending in the hawkmoth *Manduca sexta, Journal of Experimental Biology* **206**, 2999–3006.

Cooter, R. J. and Baker, P. S. (1977). Weis-Fogh clap and fling mechanism in locusta, *Nature* (*London*) **269**, 53–4.

Cox, J. (1973). The revolutionary Kasper wing, *Soaring*, December, 20.

Crabtree, L. F. (1957). Effect of leading edge separation on thin wings in two-dimensional incompressible flow, *Journal of Aeronautical Sciences* **24**, 597–604.

Cummings, R. M., Morton, S. A., Siegel, S. G., and Bosscher, S. (2003). Numerical prediction and wind tunnel experiment for pitching unmanned combat air vehicles, *AIAA Paper 2003-0417*.

Davis, R. L., Carter, J. E., and Reshotko, E. (1987). Analysis of transitional separation bubbles on infinite swept wings, *AIAA Journal* **25**, 421–8.

Davis, W. R., Kosicki, B. B., Boroson, D. M., and Kostishack, D. F. (1996). Micro air vehicles for optical surveillance, *Lincoln Laboratory Journal* **9**, 197–214.

DeLaurier, J. D. (1993). An aerodynamic model for flapping wing flight, *Aeronautical Journal* **97**, 125–130.

de Matteis, G. and de Socio, L. (1986). Nonlinear aerodynamics of a two-dimensional membrane airfoil with separation, *Journal of Aircraft* **23**, 831–6.

Devin, S. I., Zavyalov, V. M., and Korovich, B. K. (1972). On the question of unsteady aerodynamic forces acting upon a wing of finite aspect ratio at large amplitudes of oscillation and large Strouhal numbers, *Voprosy Sudostroeniya Ser.: Proektirovanie Sudov, Vyp.* **1**, 34–41.

De Vries, O. (1983). On the theory of the horizontal-axis wind turbine, *Annual Review of Fluid Mechanics* **15**, 77–96.

Dhawan, S. (1991). Bird flight, *Sadhana – Academy Proceedings in Engineering Sciences* **16**, 275–352.

Dial, K. P. (1994). An inside look at how birds fly: Experimental studies of the internal and external processes controlling flight, *1994 Report to the Aerospace Profession, 38th Symposium Proceedings*, Beverly Hills, CA.

Dick, E. and Steelant, J. (1996). Modeling of bypass transition with conditioned Navier–Stokes equations coupled to an intermittency transport equation, *International Journal for Numerical Methods in Fluids* **23**, 193–220.

Dick, E. and Steelant, J. (1997). Coupled solution of the steady compressible Navier–Stokes equations and the k–ε turbulence equations with a multigrid method, *Applied Numerical Mathematics* **23**, 49–61.

Dickinson, M. H. and Gotz, K. G. (1993). Unsteady aerodynamic perfornamce of model wings at low Reynolds numbers, *Journal of Experimental Biology* **174**, 45–64.

Dickinson, M. H., Lehmann, F.-O., and Sane, S. P. (1999). Wing rotation and the aerodynamic basis of insect flight, *Science* **284**, 1954–60.

Ding, H., Yang, B., Lou, M., and Fang, H. (2003). New numerical method for two-dimensional partially wrinkled membranes, *AIAA Journal* **41**, 125–32.

Dong, H., Mittal, R., and Najjar, F. M. (2006). Wake topology and hydrodynamic performance of low-aspect-ratio flapping foils, *Journal of Fluid Mechanics* **566**, 309–43.

Drela, M. (1989). XFOIL: An analysis and design system for low Reynolds number airfoils, in T. J. Mueller (Ed.), *Proceedings of the Conference on Low Reynolds Number Aerodynamics* (Notre Dame, University of Notre Dame Press), pp. 1–12.

Dudley, R. (2000). *The Biomechanics of Insect Flight: Form, Function, Evolution* (Princeton, NJ, Princeton University Press).

Dudley, R. and Ellington, C. P. (1990a). Mechanics of forward flight in bumblebees. I. Kinematics and morphology, *Journal of Experimental Biology* **148**, 19–52.

Dudley, R. and Ellington, C. P. (1990b). Mechanics of forward flight in bumblebees. II. Quasi-steady lift and power requirements, *Journal of Experimental Biology* **148**, 53–88.

Ellington, C. P. (1984a). The aerodynamics of hovering insect flight. I. The quasi-steady analysis, *Philosophical Transactions of the Royal Society of London. Series B* **305**, 1–15.

Ellington, C. P. (1984b). Morphological parameters, II. The aerodynamics of hovering insect flight, *Philosophical Transactions of the Royal Society of London. Series B* **305**, 17–40.

Ellington, C. P. (1984c). The aerodynamics of insect flight. III. Kinematics, *Philosophical Transactions of the Royal Society of London. Series B* **305**, 41–78.

Ellington, C. P. (1984d). The aerodynamics of hovering insect flight. IV. Aerodynamic mechanisms, *Philosophical Transactions of the Royal Society of London. Series B* **305**, 79–113.

Ellington, C. P. (1984e). The aerodynamics of hovering insect flight. V. A Vortex theory, *Philosophical Transactions of the Royal Society of London. Series B* **305**, 115–44.

Ellington, C. P. (1984f). The aerodynamics of hovering insect flight. VI. Lift and power require-ments, *Philosophical Transactions of the Royal Society of London. Series B* **305**, 145–181.

Ellington, C. P. (1995). Unsteady aerodynamics of insect flight, in C. P. Ellington and T. J. Pedley (Eds.), *Biological Fluid Dynamics*, Society for Experimental Biology Symposium, Vol. 49 (Cambridge, UK, The Company of Biologists), pp. 109–29.

Ellington, C. P., Van den Berg, C., Willmott, A. P., and Thomas, A. L. R. (1996). Leading-edge vortices in insect flight, *Nature (London)* **384**, 626–30.

Ennos, A. R. (1989). The kinematics and aerodynamics of the free flight of some *Diptera*, *Journal of Experimental Biology* **142**, 49–85.

Erickson, G. E. and Campbell, J. F. (1975). Flow visualization of leading-edge vortex enhance-ment by spanwise blowing, *NASA TM X-72702*.

Escudier, M. (1988). Vortex breakdown: Observations and explanations, *Progress in Aerospace Sciences* **25**, 189–229.

Freymuth, P. (1988). Propulsive vortical signatures of plunging and pitching airfoils, *AIAA Paper 88-323*.

Freymuth, P. (1990). Thrust generation by an airfoil in hover modes, *Experiments in Fluids* **9**, 17–24.

Friedmann, P. P. (1999). Renaissance of aeroelasticity and its future, *Journal of Aircraft* **36**, 105–21.

Fry, S. N., Sayaman, R., and Dickinson, M. H. (2003). The aerodynamics of free-flight maneu-vers in *Drosophila, Science* **300**, 495–8.

Fung, Y. C. (1969). *An Introduction to the Theory of Aeroelasticity* (New York, Dover).

Galvao, R., Israeli, E., Song, A., Tian, X., Bishop, K., Swartz, S., and Breuer, K. (2006). The aerodynamics of compliant membrane wings modeled on mammalian flight mechanics, *AIAA Paper 2006-2866*.

Garcia, H., Abdulrahim, M., and Lind, R. (2003). Roll control for a micro air vehicle using active wing morphing, *AIAA Paper 2003-5347*.

Gleyzes, C., Cousteix, J., and Bonnet, J. L. (1985). Theoretical and experimental study of low Reynolds number transitional separation bubbles, in T. J. Mueller (Ed.), *Proceedings of the Conference on Low Reynolds Number Airfoil Aerodynamics* (Notre Dame, IN, University of Notre Dame Press), pp. 137–52.

Goldspink, G. (1977). Energy cost of locomotion, in R. M. Alexander and G. C. Chapman (Eds.), *Mechanics and Energetics of Animal Locomotion* (London, Chapman and Hall).

Gopalkrishnan, R., Triantafyllou, M. S., Triantafyllou, G. S., and Barrett, D. (1994). Active vorticity control in a shear flow using a flapping foil, *Journal of Fluid Mechanics* **274** (Sep.), 1–21.

Goslow Jr., G. E., Dial, K. P., and Jenkins Jr., F. A. (1990). Bird flight: Insights and complica-tions, *BioScience* **40**, 108–15.

Gotz, K. G. (1987). Course-control, metabolism and wing interference during ultralong teth-ered flight in *Drosophila melanogaster, Journal of Experimental Biology* **128**, 35–46.

Green, A. E. and Adkins, J. E. (1960). *Large Elastic Deformations* (Oxford, Clarendon).

Greenewalt, C. H. (1975). The flight of birds: The significant dimensions, their departure from the requirements for dimensional similarity, and the effect on flight aerodynamics of that departure, *Transactions of the American Philosophical Society* **65** (4), 1–67.

Greenhalgh, S., Curtiss, H. C., and Smith, B. (1984). Aerodynamic properties of a two-dimensional inextensible flexible airfoil, *AIAA Journal* **22**, 865–70.

Grodnitsky, D. L. (1999). *Form and function of insect wings: The evolution of biological struc-tures* (Baltimore, MD, Johns Hopkins University Press).

Hall, M. G. (1972). Vortex breakdown, *Annual Review of Fluid Mechanics* **4**, 195–218.

Ham, N. D. (1968). Aerodynamic loading on a two-dimensional airfoil during dynamic stall, *AIAA Journal* **6**, 1927–34.

Harper, P. W. and Flanigan, R. E. (1950). The effect of rate of change of angle of attack on the maximum lift of a small model, *NACA TN-2061*.

Harris, F. D. and Pruyn, R. R. (1968). Blade stall–Half fact, half fiction, *Journal of the American Helicopter Society* **13**(2), 27–48.

He, X., Senocak, I., Shyy, W., Thakur, S. S., and Gangadharan, S. (2000). Evaluation of laminar-turbulent transition and equilibrium near wall turbulence models, *Numerical Heat Transfer, Part A* **37**, 101–12.

Heathcote, S., Martin, D., and Gursul, I. (2004). Flexible flapping airfoil propulsion at zero freestream velocity, *AIAA Journal* **42**, 2196–204.

Herbert, T. (1997). Parabolized stability equations, *Annual Review of Fluid Mechanics* **29**, 245–83.

Hill, A. V. (1950). The dimensions of animals and their muscular dynamics, *Science Progress* **38**, 209–30.

Hillier, R. and Cherry, N. J. (1981). The effects of stream turbulence on separation bubbles, *Journal of Wind Engineering and Industrial Aerodynamics* **8**, 49–58.

Ho, S., Nassef, H., Pornsinsirirak, N., Tai, Y.-C., and Ho, C.-M. (2003). Unsteady aerodynamics and flow control for flapping wing flyers, *Progress in Aerospace Sciences* **39**, 635–81.

Hoff, W. (1919). Der Flug der Insekten, *Naturwissenschaften* **7**, 159.

Holloway, D. S., Walters, D. K., and Leylek, J. H. (2004). Prediction of unsteady, separated boundary layer over a blunt body for laminar, turbulent, and transitional flow, *International Journal for Numerical Methods in Fluids* **45**, 1291–1315.

Houghton, E. L. and Carpenter, P. W. (2003). *Aerodynamics for engineering students* (Burlington, MA, Butterworth-Heinemann).

Hsiao, F.-B., Liu, C.-F., and Tang, Z. (1989). Aerodynamic performance and flow structure studies of a low Reynolds number airfoil, *AIAA Journal* **27**, 129–37.

Huang, R. F., Shy, W. W., Lin, S. W., and Hsiao, F.-B. (1996). Influence of surface flow on aerodynamic loads of a cantilever wing, *AIAA Journal* **34**, 527–32.

Hurley, D. G. (1959). The use of boundary-layer control to establish free stream-line flows, *Advances in Aeronautical Science* **2**, 662–708.

Ifju, P. G., Jenkins, A. D., Ettingers, S., Lian, Y., and Shyy, W. (2002). Flexible-wing-based micro air vehicles, *AIAA Paper 2002-0705*.

Isogai, K., Fujishiro, S., Saitoh, T., Yamamoto, M., Yamasaki, M., and Matsubara, M. (2004). Unsteady three-dimensional viscous flow simulation of a dragonfly hovering, *AIAA Journal* **42**, 2053–2059.

Jackson, P. (2001). *Jane's All the World's Aircraft*, (Alexandria, VA, Jane's Information Group).

Jackson, P. S. (1983). A simple model for elastic two-dimensional sails, *AIAA Journal* **21**, 153–5.

Jackson, P. S. and Christie, G. W. (1987). Numerical analysis of three-dimensional elastic membrane wings, *AIAA Journal* **25**, 676–82.

Jenkins, C. H. (1996). Nonlinear dynamic response of membranes: State of the art–update, *Applied Mechanics Reviews* **49**, S41-S48.

Jenkins, C. H. and Leonard, J. W. (1991). Nonlinear dynamic response of membranes: State of the art, *Applied Mechanics Reviews* **44**, 319–28.

Jones, B. M. (1938). Stalling, *Journal of the Royal Aeronautical Society* **38**, 747–70.

Jones, K. D., Dohring, C. M., and Platzer, F. M. (1998). Experimental and computational investigation of the Knoller–Betz effect, *AIAA Journal* **36**, 1240–6.

Jones, K. D., Lund, T. C., and Platzer, F. M. (2001). Experimental and computational investigation of flapping-wing propulsion for micro air vehicles, in T. J. Mueller (Ed.), *Fixed and Flapping Wings Aerodynamics for Micro Air Vehicle Applications*, Progress in Astronautics and Aeronautics, Vol. 195 (Reston, VA, AIAA), pp. 307–36.

Jones, K. D. and Platzer, F. M. (2006). Bio-inspired design of flapping-wing micro air vehicles – An engineer's perspective, *AIAA Paper 2006-0037*.

Jones, K. D. and Platzer, M. F. (1999). An experimental and numerical investigation of flapping-wing propulsion, *AIAA Paper 1999-0995*.

Jones, K. D. and Platzer, M. F. (2003). Experimental investigation of the aerodynamic characteristics of flapping-wing micro air vehicles, *AIAA Paper 2003-0418*.

Jones, R. T. (1990). *Wing Theory* (Princeton, NJ, Princeton University Press).

Kasper, W. (1979). *The Kasper Wing*, H. J. Meheen (Ed.), (Denver, CO, Meheen Engineering).

Katz, J. (1979). *Low-Speed Aerodynamics: From Wing Theory to Panel Methods* (San Francisco, CA, McGraw-Hill).

Katz, J. and Plotkin, A. (2002). *Low-Speed Aerodynamics* (Cambridge, UK, Cambridge University Press).

Katzmayr, R. (1922). Effect of periodic changes of angle of attack on behavior of airfoils, *NACA TM-147*.

Kawamura, Y., Souda, S., and Ellington, C. P. (2006). Quasi-hovering flight of a flapping MAV with large angle of attack, presented at *The Third International Symposium on Aero Aqua Bio-Mechanisms*, Okinawa Convention Center, Ginowan, Okinawa, Japan.

Kesel, A. B. (1998). Biologisches Vorbild Insektenflügel Mehrkriterienoptimierung ultraleichter Tragflächen, in W. Nachtigall and A. Wisser (Eds.), *Biona-Report*, Vol. 12 (Stuttgart/ New York, Fischer), pp. 107–17.

Kesel, A. B. (2000). Aerodynamic characteristics of dragonfly wing sections compared with technical airfoils, *Journal of Experimental Biology* **203**, 3125–35.

Kirkpatrick, S. J. (1994). Scale effects on the stresses and safety factors in the wing bones of birds and bats, *Journal of Experimental Biology* **190**, 195–215.

Kiya, M. and Sasaki, K. (1983). Free-stream turbulence effects on a separation bubble, *Journal of Wind Engineering and Industrial Aerodynamics* **14**(1–3), 375–86.

Knoller, R. (1909). Die Gesetze des Luftwiderstandes, *Flug-und Motortechnik (Wein)* **3**(21), 1–7.

Koochesfahani, M. M. (1989). Vortical patterns in the wake of an oscillating airfoil, *AIAA Journal* **27**, 1200–5.

Kramer, M. (1932). Die Zunahme des Maximalauftriebes von Tragflügeln bei plötzlicher Anstellwinkelvergrösserung (Böeneffect), *Zeitschrift für Flugtechnik und Motorluftschiffahrt* **23**(7), 185–9.

Kruppa, E. W. (1977). A wind tunnel investigation of the Kasper vortex concept, *AIAA Paper 77-310*.

Lai, C. S. J. and Platzer, F. M. (1999). Jet characteristics of a plunging airfoil, *AIAA Journal* **37**, 1529–37.

Lai, C. S. J. and Platzer, F. M. (2001). Characteristics of a plunging airfoil at zero freestream velocity, *AIAA Journal* **39**, 531–4.

LaRoche, U. and Palffy, S. (1996). Wing grid, a novel device for reduction of induced drag on wings, presented at the *International Council of Aeronautical Sciences (ICAS) Conference*, Sorrento, Italy.

Lehmann, F.-O. (2004). The mechanisms of lift enhancement in insect flight, *Naturwissenschaften* **91**(3), 101–22.

Lehmann, F.-O. and Dickinson, M. H. (1998). The control of wing kinematics and flight forces in fruit flies (*Drosophila spp.*), *Journal of Experimental Biology* **201**, 385–401.

Lehmann, F.-O., Sane, S. P., and Dickinson, M. H. (2005). The aerodynamic effects of wing–wing interaction in flapping insect wings, *Journal of Experimental Biology* **208**, 3075–92.

Leibovich, S. (1978). The structure of vortex breakdown, *Annual Review of Fluid Mechanics* **10**, 221–46.

Lesieur, M. and Metais, O. (1996). New trends in large-eddy simulations of turbulence, *Annual Review of Fluid Mechanics* **28**, 45–82.

Lian, Y. (2003). *Membrane and Adaptively-Shaped Wings for Micro Air Vehicles*, Ph.D. dissertation, Mechanical and Aerospace Engineering Department (Gainesville, FL, University of Florida).

Lian, Y. and Shyy, W. (2003). Three-dimensional fluid–structure interactions of a membrane wing for micro air vehicle applications, *AIAA Paper 2003-1726*.

Lian, Y. and Shyy, W. (2005). Numerical simulations of membrane wing aerodynamics for micro air vehicle applications, *Journal of Aircraft* **42**, 865–73.

Lian, Y. and Shyy, W. (2006). Laminar-turbulent transition of a low Reynolds number rigid or flexible airfoil, *AIAA Paper 2006-3051*, also AIAA Journal **45**, (2007) 1501–1513.

Lian, Y., Shyy, W., Ifju, P., and Verron, E. (2003a). A membrane wing model for micro air vehicles, *AIAA Journal* **41**, 2492–4.

Lian, Y., Shyy, W., Viieru, D., and Zhang, B. N. (2003b). Membrane wing aerodynamics for micro air vehicles, *Progress in Aerospace Sciences* **39**, 425–65.

Liebeck, R. H. (1992). Laminar separation bubbles and airfoil design at low Reynolds numbers, *AIAA Paper 1992-2735*.

Lighthill, M. J. (1969). *Hydrodynamics of Aquatic Animal Propulsion* (Philadelphia, PA, Society for Industry and Applied Mathematics).

Lighthill, M. J. (1973). On the Weis-Fogh mechanism of lift generation, *Journal of Fluid Mechanics* **60**, 1–17.

Lighthill, M. J. (1977). Introduction to the scaling of aerial locomotion, in T. J. Pedley (Ed.), *Scale Effects in Animal Locomotion* (New York, Academic), pp. 365–404.

Lissaman, P. B. S. (1983). Low Reynolds number airfoils, *Annual Review of Fluid Mechanics* **15**, 223–39.

Liu, H. (2005). Simulation-based biological fluid dynamics in animal locomotion, *Applied Mechanics Reviews* **58**, 269–282.

Liu, H., Ellington, C. P., Kawachi, K., Van den Berg, C., and Willmott, A. P. (1998). A computational fluid dynamics study of hawkmoth hovering, *Journal of Experimental Biology* **201**, 461–77.

Liu, H. and Kawachi, K. (1998). A numerical study of insect flight, *Journal of Computational Physics* **146**, 124–56.

Liu, T. (2006). Comparative scaling of flapping- and fixed-wing flyers, *AIAA Journal* **44**, 24–33.

Livne, E. (2003). Future of airplane aeroelasticity, *Journal of Aircraft* **40**, 1066–92.

Mack, L. M. (1977). Transition prediction and linear stability theory, in *Laminar-Turbulent Transition*, AGARD CP 224, pp. 1/1–22.

Maddock, L., Bone, Q., and Rayner, J. M. V. (1994). *Mechanics and Physiology of Animal Swimming* (New York, Cambridge University Press).

Malik, M. R. (1982). COSAL – A black-box compressible stability analysis code for transition prediction in three-dimensional boundary layers, *NASA CR-165925*.

Marden, J. (1987). Maximum lift production during takeoff in flying animals, *Journal of Experimental Biology* **130**, 235–58.

Mary, I. and Sagaut, P. (2002). Large eddy simulation of flow around an airfoil near stall, *AIAA Journal* **40**, 1139–45.

Maxworthy, T. (1979). Experiments on the Weis-Fogh mechanism of lift generation by insects in hovering flight. Part 1. Dynamics of the 'fling,' *Journal of Fluid Mechanics* **93**, 47–63.

Mayle, R. E. (1991). The role of laminar-turbulent transition in gas turbine engine, *Journal of Turbomachinery* **113**, 509–37.

McCroskey, W. J., Carr, L. W., and McAlister, K. W. (1976). Dynamic stall experiments on oscillating airfoils, *AIAA Journal* **14**, 57–63.

McCroskey, W. J. and Fisher, R. K. (1972). Detailed aerodynamic measurements on a model rotor in the blade stall regime, *Journal of the American Helicopter Society* **17**, 20–30.

McCroskey, W. J., McAlister, K. W., Carr, L. W., and Pucci, S. L. (1982). An experimental study of dynamic stall on advanced airfoil section, *NASA TM-84245*.

McMasters, J. H. and Henderson, M. J. (1980). Low speed single element airfoil synthesis, *Technical Soaring* **6**(2), 1–21.

McMichael, J. M. and Francis, M. S. (1997). Micro air vehicles – Toward a new dimension in flight, available at http://euler.aero.iitb.ac.in/docs/MAV/www.darpa.mil/tto/MAV/mav_auvsi.html.

Moin, P. and Mahesh, K. (1998). Direct numerical simulation: A tool in turbulence research, *Annual Review of Fluid Mechanics* **30**, 539–578.

Mooney, M. (1940). A theory of large elastic deformation, *Journal of Applied Physics* **11**, 582–592.

Mueller, T. J. (Ed.), (2001). *Fixed and Flapping Wing Aerodynamics for Micro Air Vehicle Applications*, Progress in Astronautics and Aeronautics, Vol. 195 (Reston, VA, AIAA).

Mueller, T. J. and DeLaurier, J. D. (2003). Aerodynamics of small vehicles, *Annual Review of Fluid Mechanics* **35**, 89–111.

Mueller, T. J., Pohlen, L. J., Conigliaro, P. E., and Jansen, B. J. J. (1983). The influence of free-stream disturbances on low Reynolds number airfoil experiments, *Experiments in Fluids* **1**, 3–14.

Murai, H. and Maruyama, S. (1980). Theoretical investigation of the aerodynamics of double membrane sailwing airfoil sections, *Journal of Aircraft* **17**, 294–9.

Murata, S. and Tanaka, S. (1989). Aerodynamic characteristics of a two-dimensional porous sail, *Journal of Fluid Mechanics* **206**, 463–75.

Newman, B. G. (1987). Aerodynamic theory for membranes and sails, *Progress in Aerospace Sciences* **24**, 1–27.

Newman, B. G. and Low, H. T. (1984). Two-dimensional impervious sails: Experimental results compared with theory, *Journal of Fluid Mechanics* **144**, 445–62.

Newman, B. G., Savage, S. B., and Schouella, D. (1977). Model test on a wing section of a dragonfly, in T. J. Pedley (Ed.), *Scale Effects in Animal Locomotion* (London, Academic), pp. 445–77.

Nielsen, J. N. (1963). Theory of flexible aerodynamic surfaces, *Journal of Applied Mechanics* **30**, 435–42.

Norberg, U. M. (1975). Hovering flight of the dragonfly *Aeschna juncea L.*, in T. Y.-T. Wu, C. J. Brokaw, and C. Brennen (Eds.), *Swimming and Flying in Nature*, Vol. 2 (New York, Plenum), pp. 763–81.

Norberg, U. M. (1976). Aerodynamics, kinematics, and energetics of horizontal flapping flight in the long-eared bat *Plecotus Auritus*, *Journal of Experimental Biology* **65**, 179–212.

Norberg, U. M. (1990). *Vertebrate Flight: Mechanics, Physiology, Morphology, Ecology and Evolution* (Berlin, Springer-Verlag).

Obremski, H. J. and Fejer, A. A. (1967). Transition in oscillating boundary layer flow, *Journal of Fluid Mechanics* **29**, 93–111.

Obremski, H. J. and Morkovin, M. V. (1969). Application of a quasi-steady stability model to periodic boundary layer flows, *AIAA Journal* **7**, 1298–1301.

Oden, J. T. and Sato, T. (1967). Finite strains and displacements of elastic membrane by the finite element method, *International Journal for Solids and Structures* **3**, 471–88.

Okamoto, M., Yasuda, K., and Azuma, A. (1996). Aerodynamic characteristics of the wings and body of a dragonfly, *Journal of Experimental Biology* **199**, 281–94.

Ol, M., McAuliffe, B. R., Hanff, E. S., Scholz, U., and Kaehler, C. (2005). Comparison of laminar separation bubble measurements on a low Reynolds number airfoil in three facilities, *AIAA Paper 2005-5149*.

O'Meara, M. M. and Mueller, T. J. (1987). Laminar separation bubble characteristics on an airfoil at low Reynolds numbers, *AIAA Journal* **25**, 1033–41.

Osborne, M. F. M. (1951). Aerodynamics of flapping flight with application to insects, *Journal of Experimental Biology* **28**, 221–45.

Pedley, T. J. (Ed.) (1977). *Scale Effects in Animal Locomotion* (New York, Academic).

Pendersen, C. B. and Zbikowski, R. (2006). An indicial-Polhamus aerodynamic model of insect-like flapping wings in hover, in R. Liebe (Ed.), *Flow Phenomena in Nature*, Vol. 2 (Southampton, UK, WIT Press), pp. 606–65.

Pennycuick, C. J. (1969). The mechanics of bird migration, *Ibis* **111**, 525–56.

Pennycuick, C. J. (1975). *Mechanics of Flight*, Avian Biology, D. S. Farner and J. R. King (Eds.), Vol. 5 (London, Academic).

Pennycuick, C. J. (1986). Mechanical constraints on the evolution of flight, in K. Padian (Ed.), *The Origin of Birds And the Evolution of Flight*, Memoirs of the California Academy of Sciences, Vol. 8 (San Francisco, CA, California Academy of Sciences), pp. 83–98.

Pennycuick, C. J. (1989). *Bird Flight Performance: A Practical Calculation Manual* (Oxford, UK/New York, Oxford University Press).

Pennycuick, C. J. (1990). Predicting wingbeat frequency and wavelength of birds, *Journal of Experimental Biology* **150**, 171–85.

Pennycuick, C. J. (1992). *Newton Rules Biology: A Physical Approach to Biological Problems* (New York, Oxford University Press).

Pennycuick, C. J. (1996). Wingbeat frequency of birds in steady cruising flight: New data and improved predictions, *Journal of Experimental Biology* **199**, 1613–18.

Pennycuick, C. J., Klaassen, M., Kvist, A., and Lindstrom, A. (1996). Wingbeat frequency and the body drag anomaly: Wind-tunnel observations on a thrush nightingale (*Luscinia Luscinia*) and a teal (*Anas Crecca*), *Journal of Experimental Biology* **199**, 2757–65.

Polonskiy, Y. E. (1948). Vortex streets, their application to the theory of flapping wing, Dissertatsiyay k.t.n, Moscow, VVA KA im. N. E. Zhukovskogo.

Polonskiy, Y. E. (1950). Some questions on the flapping wing, *Inzhenerniy Sbornik* **8**, 49–60.

Praisner, T. J. and Clark, J. P. (2004). Predicting transition in turbomachinery, Part I-A, Review and new model development, *ASME Paper GT2004-54108*.

Prandtl, L. and Tietjens, O. G. (1957). *Fundamentals of Hydro and Aeromechanics* (New York, Dover).

Radespiel, R., Graage, K., and Brodersen, O. (1991). Transition predictions using Reynolds-averaged Navier–Stokes and linear stability analysis methods, *AIAA Paper 91-1641*.

Radespiel, R., Windte, J., and Scholz, U. (2006). Numerical and experimental flow analysis of moving airfoils with laminar separation bubbles, *AIAA Paper 2006-501*.

Ramamurti, R. and Sandberg, W. (2001). Simulation of flow about flapping airfoils using finite element incompressible flow solver, *AIAA Journal* **39**, 253–260.

Raney, D. L. and Slominski, E. C. (2004). Mechanization and control concepts for biologically inspired micro air vehicles, *Journal of Aircraft* **41**, 1257–65.

Rayner, J. M. V. (1979a). A new approach to animal flight mechanics, *Journal of Experimental Biology* **80**, 17–54.

Rayner, J. M. V. (1979b). A vortex theory of animal flight. Part 1. The vortex wake of a hovering animal, *Journal of Fluid Mechanics* **91**, 697–730.

Rayner, J. M. V. (1979c). A vortex theory of animal flight. Part 2. The forward flight of birds, *Journal of Fluid Mechanics* **91**, 731–63.

Rayner, J. M. V. (1988). Form and function in avian flight, in R. F. Johnston (Ed.), *Current Ornithology*, Vol. 5 (New York, Plenum), pp. 1–66.

Roberts, S. K. and Yaras, M. I. (2005). Effects of surface roughness geometry on separation bubble transition, *ASME Paper GT2005-68664*.

Roberts, W. B. (1980). Calculation of laminar separation bubbles and their effect on airfoil performance, *AIAA Journal* **18**, 25–31.

Rosen, M. (1959). Water flow about a swimming fish, U.S. Navy Ordnance Test Station, *NAVWEPS Technical Report No. 2298*.

Rozhdestvensky, K. V. and Ryzhov, V. A. (2003). Aerohydrodynamics of flapping-wing propulsors, *Progress in Aerospace Sciences* **39**, 585–633.

Sane, S. P. and Dickinson, M. H. (2001). The control of flight force by a flapping wing: Lift and drag production, *Journal of Experimental Biology* **204**, 2607–26.

Sane, S. P. and Dickinson, M. H. (2002). The aerodynamic effects of wing rotation and a revised quasi-steady model of flapping flight, *Journal of Experimental Biology* **205**, 1087–96.

Satyanarayana, B. and Davis, S. (1978). Experimental studies of unsteady trailing-edge conditions, *AIAA Journal* **16**, 125–9.

Schmidt-Nielsen, K. (1984). *Scaling: Why Is Animal Size So Important?* (New York, Cambridge University Press).

Schmitz, F. W. (1942). *Aerodynamik des Flugmodells* (Berlin, Verlag).

Schrauf, G. (1998). A compressible stability code. User's Guide and Tutorial, Daimler Benz Aerospace Airbus GmbH, *Technical Report EF 040/98*.

Selig, M. S., Guglielmo, J. J., Broeren, A. P., and Giguere, P. (1995). *Summary of Low-Speed Airfoil Data*, Vol. 1 (Virginia Beach, VA, SoarTech Publications).

Selig, M. S., Guglielmo, J. J., Broeren, A. P., and Giguere, P. (1996a). Experiments on airfoils at low Reynolds numbers, *AIAA Paper 1996-0062*.

Selig, M. S., Lyon, C. A., Giguere, P., Ninham, C. N., and Guglielmo, J. J. (1996b). *Summary of Low-Speed Airfoil Data*, Vol. 2 (Virginia Beach, VA, SoarTech Publications).

Selig, M. S. and Maughmer, M. D. (1992). Multipoint inverse airfoil design method based on conformal mapping, *AIAA Journal* **30**, 1162–1170.

Shevell, R. S. (1983). *Fundamentals of Flight* (Englewood Cliffs, NJ, Prentice-Hall).

Shipman, P. (1998). *Taking Wing: Archaeopteryx and the Evolution of Bird Flight* (New York, Simon and Schuster).

Shyy, W., Berg, M., and Ljungqvist, D. (1999a). Flapping and flexible wings for biological and micro vehicles, *Progress in Aerospace Sciences* **35**, 455–506.

Shyy, W., Jenkins, D. A., and Smith, R. W. (1997). Study of adaptive shape airfoils at low Reynolds number in oscillatory flow, *AIAA Journal* **35**, 1545–48.

Shyy, W., Kleverbring, F., Nilsson, M., Sloan, J., Carroll, B., and Fuentes, C. (1999b). Rigid and flexible low Reynolds number airfoils, *Journal of Aircraft* **36**, 523–9.

Shyy, W. and Liu, H. (2007). Flapping wings and aerodynamic lift: the role of leading-edge vortices, to appear in AIAA Journal.

Shyy, W. and Smith, R. (1997). A study of flexible airfoil aerodynamics with application to micro aerial vehicles, *AIAA Paper 97-1933*.

Shyy, W., Udaykumar, H. S., Madhukar, M. R., and Richard, W. S. (1996). *Computational Fluid Dynamics with Moving Boundaries*, Series in Computational and Physical Processes in Mechanics and Thermal Sciences (Washington, D.C., Taylor and Francis).

Singh, B. and Chopra, I. (2006). Dynamics of insect-based flapping wings: Loads validation, *AIAA Paper 2006-1663*.

Singh, R. K., Chao, J., Popescu, M., Tai, C.-F., Mei, R., and Shyy, W. (2006). Multiphase/multidomain computations using continuum conservative and lattice Boltzmann methods, *ASCE Journal of Aerospace Engineering* **19**, 288–95.

Singh, R. K. and Shyy, W. (2006). Three-dimensional adaptive grid computation with conservative, marker-based tracking for interfacial fluid dynamics, *AIAA Paper 2006-1523*.

Smith, A. M. O. and Gamberoni, N. (1956). Transition, pressure gradient, and stability theory, Douglas Aircraft Co., *Report No. ES 26388*.

Smith, M. J. C. (1996). Simulating moth wing aerodynamics: Towards the development of flapping-wing technology, *AIAA Journal* **34**, 1348–55.

Sneyd, A. D. (1984). Aerodynamic coefficients and longitudinal stability of sail airfoils, *Journal of Fluid Mechanics* **149**, 127–46.

Spedding, G. R. (1992). The aerodynamics of flight, in R. M. Alexander (Ed.), *Mechanics of Animal Locomotion*, Advances in Comparative and Environmental Physiology, Vol. 11 (Berlin, Springer-Verlag), pp. 52–111.

Srygley, R. B. and Thomas, A. L. R. (2002). Unconventional lift-generating mechanisms in free-flying butterflies, *Nature (London)* **420**, 660–4.

Stanford, B., Viieru, D., Albertani, R., Shyy, W., and Ifju, P. (2006). A numerical and experimental investigation of flexible micro air vehicle wing deformation, *AIAA Paper 2006-0440*.

Stock, H. W. and Haase, W. (1999). A feasibility study of e^N transition prediction in Navier–Stokes methods for airfoils, *AIAA Journal* **37**, 1187–96.

Storer, J. H. (1948). *The Flight of Birds*, Cranbrook Institute Bulletin, 28 (Bloomfield Hills, MI, Cranbrook Press).

Streitlien, K. and Triantafyllou, G. S. (1998). On thrust estimates for flapping foils, *Journal of Fluids and Structures* **12**, 47–55.

Sugimoto, T. and Sato, J. (1988). Aerodynamic characteristics of two-dimensional membrane airfoils, *Journal of the Japan Society for Aeronautical and Space Sciences* **36**, 36–43.

Sun, M. and Tang, J. (2002a). Unsteady aerodynamic force generation by a model fruit fly wing in flapping motion, *Journal of Experimental Biology* **205**, 55–70.

Sun, M. and Tang, J. (2002b). Lift and power requirements of hovering flight in *Drosophila virilis*, *Journal of Experimental Biology* **205**, 2413–27.

Sunada, S. and Ellington, C. P. (2000). Approximate added-mass method for estimationg induced power for flapping fight, *AIAA Journal* **38**, 1313–21.

Sunada, S., Kawachi, K., Matsumoto, A., and Sakaguchi, A. (2001). Unsteady forces on a two-dimensional wing in plunging and pitching motions, *AIAA Journal* **39**, 1230–9.

Sunada, S., Kawachi, K., Watanabe, I., and Azuma, A. (1993). Fundamental analysis of three-dimensional 'near fling,' *Journal of Experimental Biology* **183**, 217–48.

Sunada, S., Yasuda, T., Yasuda, K., and Kawachi, K. (2002). Comparison of wing characteristics at an ultralow Reynolds number, *Journal of Aircraft* **39**, 331–8.

Suzen, Y. B. and Huang, P. G. (2000). Modeling of flow transition using an intermittency transport equation, *Journal of Fluids Engineering* **122**, 273–84.

Swartz, S. M. (1997). Allometric patterning in the limb skeleton of bats: Implications for the mechanics and energies of powered flight, *Journal of Morphology* **234**, 277–94.

Swartz, S. M., Bennett, M. B., and Carrier, D. R. (1992). Wing bone stresses in free flying bats and the evolution of skeletal design for flight, *Nature (London)*, **359**, 726–9.

Taneda, S. (1976). Visual study of unsteady separated flows around bodies, *Progress in Aerospace Sciences* **17**, 287–348.

Tang, J., Viieru, D., and Shyy, W. (2007). Effects of Reynolds number, reduced frequency and flapping kinematics on hovering aerodynamics, *AIAA Paper 2007-0129*.

Tang, J. and Zhu, K.-Q. (2004). Numerical and experimental study of flow structure of low-aspect-ratio wing, *Journal of Aircraft* **41**, 1196–1201.

Tani, I. (1964). Low-speed flows involving bubble separations, in D. Kuchenmann and L. H. G. Sterne (Eds.), *Progress in Aeronautical Sciences*, Vol. 5 (New York, Pergamon), pp. 70–103.

Taylor, G. K., Nudds, R. L., and Thomas, A. L. R. (2003). Flying and swimming animals cruise at a Strouhal number tuned for high power efficiency, *Nature (London)* **425**, 707–11.

Templin, R. J. (2000). The spectrum of animal flight: Insects to pterosaurs, *Progress in Aerospace Sciences* **36**, 393–436.

Tennekes, H. (1996). *The Simple Science of Flight (From Insects to Jumbo Jets)* (Boston, MIT Press).

Thomas, A. L. R., Taylor, G. K., Srygley, R. B., Nudds, L. R., and Bomphrey, R. J. (2004). Dragonfly flight: Free-flight and tethered flow visualizations reveal a diverse array of unsteady lift-generating mechanisms, controlled primarily via angle of attack, *Journal of Experimental Biology* **207**, 4299–323.

Thwaites, B. (1961). The aerodynamic theory of sails. Part I. Two-dimensional sails, *Proceedings of the Royal Society of London. Series A* **261**, 402–22.

Tian, X., Iriarte, J., Middleton, K., Galvao, R., Israeli, E., Roemer, A., Sullivan, A., Song, A., Swartz, S., and Breuer, K. (2006). Direct measurements of the kinematics and dynamics of bat flight, *AIAA Paper 2006-2865*.

Tobalske, B. W. and Dial, K. P. (1996). Flight kinematics of black-billed magpies and pigeons over a wide range of speeds, *Journal of Experimental Biology* **199**, 263–80.

Tobalske, B. W., Hedrick, T. L., Dial, K. P., and Biewener, A. A. (2003). Comparative power curves in bird flight, *Nature (London)* **421**, 363–6.

Torres, G. E. and Mueller, T. J. (2001). Aerodynamic characteristics of low aspect ratio wings at low Reynolds numbers, in T. J. Mueller (Ed.), *Fixed and Flapping Wing Aerodynamics for Micro Air Vehicles*, Progress in Astronautics and Aeronautics, Vol. 195 (Reston, VA, AIAA), pp. 341–91.

Triantafyllou, M. S., Triantafyllou, G. S., and Yue, D. K. P. (2000). Hydrodynamics of fishlike swimming, *Annual Review of Fluid Mechanics* **32**, 33–53.

Usherwood, J. R. and Ellington, C. P. (2002). The aerodynamics of revolving wings I. Model hawkmoth wings, *Journal of Experimental Biology* **205**, 1547–64.

Van den Berg, C. and Ellington, C. P. (1997). The three-dimensional leading-edge vortex of a 'hovering' model hawkmoth, *Philosophical Transactions of the Royal Society of London. Series B* **352**, 329–40.

Vanden-Broeck, J. M. (1982). Nonlinear two-dimensional sail theory, *Physics of Fluids* **25**, 420–3.

Vanden-Broeck, J. M., and Keller, J. B. (1981). Shape of a sail in a flow, *Physics of Fluids* **24**, 552–3.

Van Ingen, J. L. (1956). A suggested semi-empirical method for the calculation of the boundary layer transition region, Delft University of Technology, Dept. of Aerospace Engineering, *Report No. VTH-74*.

Van Ingen, J. L. (1995). Some introductory remarks on transition prediction methods based on linear stability theory, in R. A. W. M. Henkes and J. L. van Ingen (Eds.), *Transitional Boundary Layers in Aeronautics* (Amsterdam, The Netherlands, Elsevier), pp. 209–24.

Verron, E., Marckmann, G., and Pesaux, B. (2001). Dynamic inflation of non-linear elastic and viscoelastic rubber-like membranes, *International Journal for Numerical Methods in Engineering* **50**, 1233–51.

Vest, M. S. and Katz, J. (1996). Unsteady aerodynamics model of flapping wings, *AIAA Journal* **34**, 1435–40.

Videler, J. J., Stamhuis, E. J., and Povel, G. D. E. (2004). Leading-edge vortex lifts swifts, *Science* **306**, 1960–2.

Viieru, D., Albertani, R., Shyy, W., and Ifju, G. P. (2005). Effect of tip vortex on wing aerodynamics of micro air vehicles, *Journal of Aircraft* **42**, 1530–6.

Viieru, D., Lian, Y., Shyy, W., and Ifju, G. P. (2003). Investigation of tip vortex on aerodynamic performance of a micro air vehicle, *AIAA Paper 2003-3597*.

Viieru, D., Tang, J., Lian, Y., Liu, H., and Shyy, W. (2006). Flapping and flexible wing aerodynamics of low Reynolds number flight vehicles, *AIAA Paper 2006-0503*.

Voelz, K. (1950). Profil und Luftriebeines Segels, *Zeitschrift für Angewandte Mathematik und Mechanik* **30**, 301–17.

Vogel, S. (1967). Flight in Drosophila. III. Aerodynamic characteristics of fly wings and wing models, *Journal of Experimental Biology* **46**, 431–43.

Vogel, S. (1996). *Lift in Moving Fluids: The Physical Biology of Flow* (Princeton, NJ, Princeton University Press).

Volino, R. J. and Bohl, D. G. (2004). Separated flow transition mechanism and prediction with high and low freestream turbulence under low pressure turbine conditions, *ASME Paper GT2004-53360*.

Von Karman, T. and Burgers, J. M. (1935). General aerodynamic theory – Perfect fluids, in W. Durand (Ed.), *Aerodynamic Theory*, Vol. II (Berlin, Springer).

Wakeling, J. M. and Ellington, C. P. (1997a). Dragonfly flight. II. Velocities, accelerations and kinematics of flapping flight, *Journal of Experimental Biology* **200**, 557–82.

Wakeling, J. M. and Ellington, C. P. (1997b). Dragonfly flight. III. Lift and power requirements, *Journal of Experimental Biology* **200**, 583–600.

Walker, J. A. and Westneat, M. W. (2000). Mechanical performance of aquatic rowing and flying, *Proceedings of the Royal Society of London. Series B* **267**, 1875–81.

Wang, Z. J. (2000). Vortex shedding and frequency selection in flapping flight, *Journal of Fluid Mechanics* **410**, 323–41.

Wang, Z. J., Birch, J. M., and Dickinson, M. H. (2004). Unsteady forces and flows in low Reynolds number hovering flight: Two-dimensional computations vs robotic wing experiments, *Journal of Experimental Biology* **207**, 449–60.

Ward-Smith, A. J. (1984). *Biophysical Aerodynamics and the Natural Environment* (New York, Wiley).

Warrick, D. R., Tobalske, B. W., and Powers, D. R. (2005). Aerodynamics of the hovering hummingbird, *Nature (London)* **435**, 1094–7.

Waszak, R. M., Jenkins, N. L., and Ifju, P. (2001). Stability and control properties of an aeroelastic fixed wing micro aerial vehicle, *AIAA Paper 2001-4005*.

Wazzan, A. R., Gazley, J. C., and Smith, A. M. O. (1979). Tollmien–Schlichting waves and transition: Heated and adiabatic wedge flows with application to bodies of revolution, *Progress in Aerospace Sciences* **18**, 351–92.

Wazzan, A. R., Okamura, T. T., and Smith, A. M. O. (1968). Spatial and temporal stability charts for the Falkner–Skan boundary layer profiles, Douglas Aircraft Co, *DAC-67086*.

Weis-Fogh, T. (1972). Energetics of hovering flight in hummingbirds and in drosophila, *Journal of Experimental Biology* **56**, 79–104.

Weis-Fogh, T. (1973). Quick estimates of flight fitness in hovering animals, including novel mechanisms for lift production, *Journal of Experimental Biology* **59**, 169–230.

Weis-Fogh, T. and Jensen, M. (1956). Biology and physics of locust flight. I. Basic principles in insect flight. A critical review, *Philosophical Transactions of the Royal Society of London. Series B* **239**, 415–58.

Weiss, H. (1939). Wind tunnel: Effect of wing spar size, *Journal of International Aeromodeling*, pp. 5–7.

Westesson, R. A. and Clareus, U. (1974). Turbulent lift. Comments on some preliminary wind tunnel tests – Characteristics of vortex on wing surface from tangential blowing on upper surface, *NASA-TT-F-15743, TP-74-51*.

White, F. M. (1991). *Viscous Fluid Flow* (New York, McGraw-Hill).

Wilcox, C. D. (2000). *Turbulence Modeling for CFD* (La Canada, CA, DCW Industries).

Wilcox, D. C. (1994). Simulation of transition with a two-equation turbulence model, *AIAA Journal* **32**, 247–55.

Wilkin, P. J. and Williams, H. M. (1993). Comparison of the aerodynamic forces on a flying sphingid moth with those predicted by quasi-steady theory, *Physiological Zoology* **66**, 1015–44.

Willmott, A. P. and Ellington, C. P. (1997a). Measuring the angle of attack of beating insect wings: Robust three-dimensional reconstruction from two-dimensional images, *Journal of Experimental Biology* **200**, 2693–2704.

Willmott, A. P. and Ellington, C. P. (1997b). The mechanics of flight in the hawkmoth Manduca Sexta. I. Kinematics of hovering and forward flight, *Journal of Experimental Biology* **200**, 2705–22.

Willmott, A. P. and Ellington, C. P. (1997c). The mechanics of flight in the hawkmoth Manduca Sexta. II. Aerodynamic consequences of kinematic and morphological variation, *Journal of Experimental Biology* **200**, 2723–45.

Wolfgang, M. J., Tolkoff, S. W., Techet, A. H., Barrett, D. S., Triantafyllou, M. S., Yue, D. K. P., Hover, F. S., Grosenbaugh, M. A., and McGillis, W. R. (1998). Drag reduction and turbulence control in swimming fish-like bodies, *Proceedings of the International Symposium on Seawater Drag Reduction* (Newport, RI, Naval Undersea Warfare Center).

Wootton, R. J. and Newman, D. J. S. (1979). Whitefly have the highest contraction frequencies yet recorded in non-fibrillar flight muscles, *Nature (London)* **280**, 402–3.

Wu, J. Z., Vakili, A. D., and Wu, J. M. (1991). Review of the physics of enhancing vortex lift by unsteady excitation, *Progress in Aerospace Sciences* **28**, 73–131.

Wu, T. Y.-T. (1971). Hydromechanics of swimming of fishes and cetaceans, in C.-S. Yih (Ed.), *Advances in Applied Mechanics*, Vol. 11 (New York, Academic), pp. 1–63.

Wu, T. Y.-T., Brokaw, C. J., and Brennen, C. (Eds.) (1975). *Swimming and Flying in Nature*, Vols. 1 and 2 (New York, Plenum).

Ye, T., Shyy, W., and Chung, J. C. (2001). A fixed-grid, sharp-interface, method for bubble dynamics and phase change, *Journal of Computational Physics* **174**, 781–815.

Young, A. D. and Horton, H. P. (1966). Some results of investigation of separation bubbles, *AGARD Conference Proceedings*, Vol. 4, Part 2 (London, UK, Technical Editing and Reproduction, Ltd.), pp. 785–811.

Yuan, W., Khalid, M., Windte, J., Scholz, U., and Radespiel, R. (2005). An investigation of low-Reynolds-number flows past airfoils, *AIAA Paper 2005-4607*.

Zanker, J. M. and Gotz, K. G. (1990). The wing beat of Drosophila Melanogaster. II. Dynamics, *Philosophical Transactions of the Royal Society of London. Series B* **327**, 19–44.

Zbikowski, R. (2002). On aerodynamic modelling of an insect-like flapping wing in hover for micro air vehicles, *Philosophical Transactions of the Royal Society of London. Series A* **360**, 273–90.

Zheng, X., Liu, C., Liu, F., and Yang, C. (1998). Turbulence transition simulation using the k-ω model, *International Journal for Numerical Methods in Engineering* **42**, 907–26.

Index